Lecture Notes in Mathematics　　　1570

Editors:
A. Dold, Heidelberg
B. Eckmann, Zürich
F. Takens, Groningen

Ralph deLaubenfels

Existence Families, Functional Calculi and Evolution Equations

Springer-Verlag

Berlin Heidelberg New York
London Paris Tokyo
Hong Kong Barcelona
Budapest

Author

Ralph deLaubenfels
Mathematics Department
Ohio University
315 B Morton Hall
Athens, Ohio 45701-2979, USA

Mathematics Subject Classification (1991): 47D05, 47D10, 47A60, 34G10, 35G10, 47A10, 47A12, 47F05

ISBN 3-540-57703-3 Springer-Verlag Berlin Heidelberg New York
ISBN 0-387-57703-3 Springer-Verlag New York Berlin Heidelberg

Library of Congress Cataloging-in-Publication Data. DeLaubenfels, Ralph, 1951- . Existence families, functional calculi, and evolution equations / Ralph deLaubenfels. p. cm. – (Lecture notes in mathematics; 1570) Includes bibliographical references and index. ISBN 0-387-57703-3
1. Evolution equations. 2. Linear operators. 3. Functional analysis. I. Title. II. Series: Lecture notes in mathematics (Springer-Verlag); 1570. QA3.L28 no. 1570 [QA377] 510 s–dc20 [515'.353] 93-47576

© Springer-Verlag Berlin Heidelberg 1994
Printed in Germany

SPIN: 10078796   46/3140-543210 - Printed on acid-free paper

To Karen, David, Michael
and Evan

TABLE OF CONTENTS

0. INTRODUCTION

In this book, we wish to present the basic theory, and the more interesting developments, of *existence families* for the *abstract Cauchy problem*

$$\frac{d}{dt}u(t,x) = A\left(u(t,x)\right)(t \geq 0), \ u(0,x) = x, \qquad (0.1)$$

and *regularized semigroups* (these have been recently appearing in the literature as "C-semigroups"). These are generalizations of strongly continuous semigroups, that can be applied directly to the many differential and integral equations that may be modeled as an abstract Cauchy problem on a Banach space, where strongly continuous semigroups cannot be applied directly, that is, when the operator A in (0.1) does not generate a strongly continuous semigroup. Examples of this are the backwards heat equation, the Schrödinger equation on $L^p, p \neq 2$, the Cauchy problem for the Laplace equation and Petrowsky correct systems of constant coefficient partial differential equations.

These families of operators unify well-posed and ill-posed problems, cover the entire spectrum of ill-posedness and unify and simplify results about ill-posed problems, just as strongly continuous semigroups do with well-posed problems.

Of particular interest in this book will be the interplay between functional, or operational, calculus constructions, existence families and evolution equations. We do not attempt to give detailed physical applications. The examples of evolution equations are meant primarily to illustrate the great scope and broad, yet simple and intuitive, applicability of existence families and regularized semigroups. We shall see that regularized semigroups may also be used as a powerful tool for constructing functional calculi for unbounded operators.

More complete descriptions of how physical problems may be modeled as an abstract Cauchy problem (0.1) may be found in any of the references for strongly continuous semigroups.

Another central feature of this book is what is called the *solution space*, for an arbitrary closed operator. This is a unifying concept for all the generalizations of strongly continuous semigroups that have appeared, besides the existence families presented here. Even for the strongly continuous case, the introduction of the solution space greatly simplifies the proof of the fundamental relationship between the abstract Cauchy problem and semigroups.

In this book, we have tried to avoid those aspects of C-regularized semigroups that are obvious generalizations of strongly continuous semigroups, that is, where one takes a proof for strongly continuous semigroups and places a C everywhere. It is gratifying to find that many of these results may be immediately reduced to the corresponding result for strongly continuous semigroups with the solution space. In this book, we have included primarily results about existence families and regularized semigroups that would seem surprising to anyone familiar with strongly continuous semigroups. The existence of entire C-regularized groups with unbounded generators is an example. Another example is immense simplications and clarifications, both in the statement of results and their proofs, of old results, involving, for example, the Schrödinger operator on $L^p, p \neq 2$, and Petrowsky correct systems of differential equations.

Many of the most fundamental results in this book, including Chapters II, V, VII and XII, are new, that is, have not appeared, at least in their present exposition, and are not intended to appear, in research journals. We have also organized and simplified a great deal of basic material that is currently scattered throughout many different papers.

Some familiarity with the basic results of strongly continuous semigroups would be helpful in reading, or at least appreciating, this book, but is not necessary for most of the material, since these results will be a corollary of our results here.

The theory of strongly continuous semigroups of operators has been extremely successful in dealing with the many situations that may be modeled as an abstract Cauchy problem (0.1). When A is closed, generating a strongly continuous semigroup corresponds to the abstract Cauchy problem having a unique mild solution, for all initial data x.

What should be done when the abstract Cauchy problem (0.1) is not well-posed or does not have a mild solution for all initial data? It does not seem constructive to merely say it is not well-posed, throw up our hands with despair and refuse to deal with it. Ill-posed problems arise naturally (see [Pay]); the initial formulation of (0.1) is often made out of convenience or familiarity. The supremum norm and the L^1 norm, for example, are very natural choices for Banach space norms, but in many cases, such as the Schrödinger equation, do not yield well-posedness.

There are two approaches one can take. One can look for initial data in the original space that yield solutions. Or one can renorm a subspace in such a way that A, when restricted to that space, generates a strongly continuous semigroup. We shall see that existence families and regularized semigroups provide a simple yet powerful tool for either approach. For the first approach, we shall see that, for any bounded operator C, having a mild solution of the abstract Cauchy problem, for all initial data in the image of C, corresponds to having a mild C-existence family. We present in this book an intuitive method for choosing such C and constructing the desired existence family. For the second approach, we show that a mild C-existence family gives us a straightforward method for constructing and approximating a subspace of the original space, containing all mild solutions of the abstract Cauchy problem, on which A generates a strongly continuous semigroup. We call this the *solution space* for A (see Chapters IV and V).

Besides existence and well-posedness, it is also important to have methods for representing or constructing solutions. We feel the essential concept here is an operator theoretic one of a good deal of independent interest, that of an *$\mathcal{F}$-functional calculus for A*. For bounded operators, this is a continuous algebra homomorphism, from the Banach algebra of functions $\mathcal{F}$ into the Banach algebra of bounded linear operators, mapping the constant function $f_0(s) \equiv 1$ to the identity operator, and mapping $f_1(s) \equiv s$ to A. Some modifications are required to adapt this concept to unbounded operators, where the subject is relatively unformed (except for spectral operators—see [Du-S2]). Informally, one is replacing the complex variable z by A, and making sense out of $f(A)$. Again informally, the solution of the abstract Cauchy problem is $u(t, x) = e^{tA}x$, that is, $f_t(A)x$, where $f_t(z) \equiv e^{tz}$. Many of the most important results about semigroups of operators can be best understood with this perspective, and many new results, presented in this book, can be created and understood similarly.

We may summarize the functional calculus approach to evolution equations, via the abstract Cauchy problem (0.1), as follows. We think of a strongly continuous semigroup generated by A as being e^{tA}. We think of a C-regularized semigroup generated by A as being $e^{tA}g(A)$, for some g such that the complex-valued function $z \mapsto e^{tz}g(z)$ decays sufficiently rapidly on the spectrum of A; here C is $g(A)$. Finally, we think of a solution of (0.1) as being $u(t, x) \equiv e^{tA}x$. Throughout this book, we will try to make literal sense out of these exponentials.

To conclude our general introduction, we should make it clear that we do not consider regularized semigroups or existence families to be any sort of revolutionary replacement of strongly continuous semigroups. We view it more as a *completion* of the theory of strongly continuous semigroups, extending and finishing a beautiful and fundamental theory that strongly continuous semigroups began. For example, when there are solutions of the abstract Cauchy problem, it is natural to ask when they are reversible (see Chapter VIII). When $\{e^{zA}\}_{Re(z)>0}$ is a strongly continuous holomorphic semigroup of angle $\frac{\pi}{2}$, it is natural to ask in what sense we can let $Re(z) \to 0$; that is, can we make sense out of boundary values e^{isA} (see Chapter X)? As with strongly continuous semigroups, an approach that allows us to deal with bounded operators has the best chance of being successful.

Now we will give a rough chapter-by-chapter introduction.

For an arbitrary bounded operator C, and an unbounded operator A, C-existence families for A, which we think of as $e^{tA}C$, are introduced in Chapter II. We must quibble about whether we are obtaining mild solutions or strong solutions (Definition 2.1) of the abstract Cauchy problem (0.1), thus we have mild existence families, existence families and strong existence families.

We introduce regularized semigroups in Chapter III. These are essentially existence families $e^{tA}C$, where $e^{tA}C = Ce^{tA}$. This gives us an algebraic definition, $W(t)W(s) = W(t+s)C$, analogous to strongly continuous semigroups (I-regularized semigroups). To enable us to define the generator, A, we insist that C be injective (so that we may differentiate Ce^{tA} at $t = 0$, then apply C^{-1}).

In Chapter IV, for an arbitrary closed operator A on a Frechet space, we introduce its *solution space*, Z. This is the set of all initial data, x, for which the abstract Cauchy problem has a mild solution. When all solutions are unique, we may topologize Z in such a way that Z is a Frechet space, and A restricted to Z generates a strongly continuous, locally equicontinuous semigroup. This is the only chapter where X is a general Frechet space, rather than a Banach space.

By restricting ourselves to exponentially bounded mild solutions of the abstract Cauchy problem, we obtain maximal *Banach* spaces, on which a closed operator A generates a strongly continuous semigroup, in Chapter V. We present numerous methods for testing whether a vector x is in one of these exponentially bounded solution spaces, and constructing the corresponding mild solution of the abstract Cauchy problem; some of these tests may be considered "pointwise Hille-Yosida theorems."

In both Chapters IV and V, we see how regularized semigroups and existence families provide a way to approximate the solution spaces and their topologies.

Chapter VI presents a strategy opposite to that of Chapters IV and V. Rather than shrink the original space, we enlarge it, to produce a space, W, in which the original space is continuously embedded, and an operator B, that generates a strongly continuous semigroup, such that the original operator A equals the restriction of B to X.

In Chapter VII, a concept that has been used a great deal for strongly continuous semigroups, that of an *entire vector*, is used to give a simple necessary and sufficient condition for an operator A to have an entire C-existence family: the image of C must consist entirely of entire vectors. We may also construct this family with the power series for the exponential function. It should be mentioned that an entire strongly continuous group (or integrated group) occurs only in the essentially trivial case of a bounded generator.

In Chapters VIII and IX, we apply Chapter VII to two famous ill-posed problems, the backwards heat equation and the Cauchy problem for the Laplace equation, giving simple proofs of the existence of unique entire solutions, for all initial data in a dense set. More generally, we show in Chapter VIII that any parabolic problem is reversible, for all initial data in a dense set.

Chapter X considers the "boundary values" $\{e^{isA}\}_{s\in\mathbf{R}}$ of a strongly continuous bounded holomorphic semigroup $\{e^{zA}\}_{Re(z)>0}$, and gives a correspondence between the values of r for which $\{e^{isA}(1-A)^{-r}\}_{s\in\mathbf{R}}$ is a polynomially bounded $(1-A)^{-r}$-regularized semigroup and the rate of growth of $\|e^{zA}\|$, as $Re(z)$ goes to zero.

In Chapter XI, Chapter X is applied to the Schrödinger equation, with potential, on $L^p(\mathbf{R}^n)(1 \le p < \infty)$.

In Chapter XII, we use regularized semigroups to define, for $f \in C^\infty(\mathbf{R}^n)$, commuting generators $iA_1, \ldots, iA_n$ of bounded strongly continuous groups, a functional calculus $f \mapsto f(A_1, \ldots, A_n)$. We explicitly define, for each such f, with the Fourier inversion theorem, a regularized semigroup, then define $f(A_1, \ldots, A_n)$ to be the generator.

We use the construction of Chapter XII to consider, in Chapters XIII and XIV, matrices of polynomials of generators of bounded strongly continuous groups. In Chapter XIII, we consider Petrowsky correct matrices, and obtain generators of exponentially bounded $(1-|A|^2)^{-r}$-regularized semigroups, for appropriate r. In Chapter XIV, we show that *any* matrix of polynomials of generators of bounded strongly continuous groups generates an entire C-regularized group, for an explicit C with dense image.

A few more examples of regularized semigroups appear in Chapter XV.

Even when C does not commute with A, in Chapter XVI we show how we may give an algebraic definition of a *pair* of families of operators, an existence family and a uniqueness family, with a generator A. We think of these families as $W_1(t) \equiv e^{tA}C_1$ (for existence) and $W_2(t) \equiv C_2 e^{tA}$ (for uniqueness); a formal calculation leads to an algebraic definition that intertwines the two families:

$$W_2(t)W_1(s) = C_2 W_1(t+s) = W_2(t+s)C_1,$$

for all $s, t \ge 0$.

Chapter XVII gives analogues of Hille-Yosida type theorems, for C-existence families, involving the natural analogue of the resolvent, the C-*resolvent*.

In Chapter XVIII, we show that an n-times integrated semigroup, another generalization of strongly continuous semigroups that has appeared recently, corresponds to an $(r - A)^{-n}$-regularized semigroup.

Some perturbation results, both additive and multiplicative, appear in Chapter XIX.

We introduce the *type* of an operator in Chapter XX. This is an analogue of numerical range that satisfies mapping theorems similar to spectral mapping theorems.

The results that one would expect for holomorphic C-existence families appear in Chapter XXI. We do not give any details of proofs here.

In Chapter XXII we use regularized semigroups and unbounded versions of the Cauchy integral formula to define a holomorphic functional calculus, $f \mapsto f(A)$, for large classes of unbounded operators A and functions f holomorphic on an open set containing the spectrum of A. We give numerous examples, including many powerful applications to the abstract Cauchy problem. This generalizes, to unbounded operators and unbounded (they may not even be polynomially bounded) functions, the Riesz-Dunford holomorphic functional calculus.

The construction of Chapter XXII is applied, in Chapter XXIII, to characterize what we call *spectral dense solution sets*. These are subsets, V, of the complex plane, with the property that, whenever the spectrum of a densely defined operator A is contained in V, with the norm of the resolvent of A polynomially bounded outside V, then the abstract Cauchy problem (0.1) has a solution, for all initial data in a dense set.

In Chapter XXIV we show that polynomials satisfy "type mapping theorems" (see Chapter XX). This leads to a large class of operators A and polynomials p such that $p(A)$ generates an integrated semigroup, hence a regularized semigroup.

In Chapter XXV we show how we may deal with many higher order abstract Cauchy problems by factoring them into an "iterated abstract Cauchy problem," with each factor generating a regularized semigroup. We use this to show that many second order problems, including the Cauchy problem for the Laplace equation, admit equipartition of energy, in Chapter XXVI.

Terminology 0.2. We will use the following terminology throughout the book. Except in Chapter IV, where X will be a Frechet space, X will be a Banach space. All operators are linear. We will write $\mathcal{D}(A)$ for the domain of the operator A, $\sigma(A)$ for the spectrum, $\rho(A)$ for the resolvent set. The space of bounded operators from X into itself will be denoted by $B(X)$. We will write $Im(C)$ for the image of an operator C. We may make $Im(C)$ into a Banach space, $[Im(C)]$, with the norm

$$\|y\|_{[Im(C)]} \equiv \inf\{\|x\| \mid Cx = y\}.$$

When A is closed, $[\mathcal{D}(A)]$ will mean $\mathcal{D}(A)$, made into a Banach space with the graph norm

$$\|x\|_{[\mathcal{D}(A)]} \equiv \|x\| + \|Ax\| \ (x \in \mathcal{D}(A)).$$

When A generates a strongly continuous semigroup, we will write $\{e^{tA}\}_{t \geq 0}$ for that semigroup.

I. INTUITION AND ELEMENTARY EXAMPLES

One of the first differential equations students see in their first class on the subject is

$$\frac{d}{dt}u(t) = au(t), \qquad (1.1)$$

where a is a complex number. It does not take long to convince one that the solution is given by $u(t) = e^{ta}u(0)$. Soon after this, one learns to deal with a system of n constant coefficient linear differential equations by writing it as a single matrix differential equation

$$\frac{d}{dt}\vec{u}(t) = A\vec{u}(t),$$

where A is an $n \times n$ matrix and $\vec{u}$ is an n-vector whose components are the unknown functions, and again the solution is given by $\vec{u}(t) = e^{tA}\vec{u}(0)$.

This simple example is a good demonstration of what we consider to be one of the major purposes of mathematics, to *simplify and clarify*. A complicated and chaotic problem is reduced to a simple and familiar problem.

Even in finite dimensions, we see how a problem in differential equations is transformed into an interesting problem in operator theory, specifically, what is called a *functional calculus* (or operational calculus). What is meant by e^{tA}? This can be defined using the power series for the exponential function,

$$e^{tA} \equiv \sum_{k=0}^{\infty} \frac{t^k A^k}{k!}, \qquad (1.2)$$

but this still leaves the problem of how to compute it. A more useful representation may be constructed using the Jordan canonical form for a matrix. For any sufficiently differentiable complex-valued function f, define

$$f(\begin{bmatrix} a & 1 \\ 0 & a \end{bmatrix}) \equiv \begin{bmatrix} f(a) & f'(a) \\ 0 & f(a) \end{bmatrix}, \; f(\begin{bmatrix} a & 1 & 0 \\ 0 & a & 1 \\ 0 & 0 & a \end{bmatrix}) \equiv \begin{bmatrix} f(a) & f'(a) & \frac{1}{2}f''(a) \\ 0 & f(a) & f'(a) \\ 0 & 0 & f(a) \end{bmatrix},$$

etc.; it is clear how to extend this definition to matrices formed by pasting together such Jordan blocks, as in the Jordan form. When A is an $n \times n$ matrix, this definition of $f(A)$ (combined with the Jordan canonical form) defines a *functional calculus* for A, that is, $(fg)(A) = f(A)g(A)$, $f_0(A) =$

1

$I, f_1(A) = A$, where $f_0(z) \equiv 1, f_1(z) \equiv z$. When $g_t(z) \equiv e^{tz}$, then $g_t(A)$ may be shown to be the same as (1.2).

A large class of partial differential initial-value problems and mixed problems may similarly be made to look like (1.1), as an *abstract Cauchy problem* (see (0.1)), where A is an operator on a locally convex space X and $u(\cdot, x) : [0, \infty) \to X$. Again the natural thing to do would be to exponentiate A to get the solution u; $u(t, x) \equiv e^{tA}x$. But what does this mean? Let us consider another special case, but this time in infinite dimensions. Suppose iA is a self-adjoint operator on a Hilbert space, H. Then there exists a functional calculus, $f \mapsto f(A)$, defined for all bounded continuous functions on the real line, given by

$$f(iA)x \equiv \int_{\mathbf{R}} f(s) \, dE(s)x \, (x \in H),$$

where E is a projection-valued measure.

This is the *spectral theorem*. It is now clear how to exponentiate A: $e^{tA} \equiv g_t(iA)$, where $g_t(s) \equiv e^{-its}$. And in fact this turns out to define solutions of the abstract Cauchy problem (0.1), as a Fourier transform, given by

$$u(t, x) \equiv e^{tA}x \equiv \int_{\mathbf{R}} e^{-its} \, dE(s)x,$$

for x in the domain of A.

In the operator-theoretic approach to the abstract Cauchy problem, one defines a *strongly continuous semigroup generated by A*, $\{T(t)\}_{t \geq 0} \equiv \{e^{tA}\}_{t \geq 0}$, to be a family of bounded operators with the properties of the family of functions $s \mapsto e^{ts}$; $T(0) = I, T(s + t) = T(t)T(s)$, for all $s, t \geq 0$ and $\frac{d}{dt}T(t)x|_{t=0} = Ax$, for all x in the domain of A. Thus this is a functional calculus, defined for exponential functions $s \mapsto e^{ts}$, for $t \geq 0$. The most well-known characterization of generators of strongly continuous semigroups, the *Hille-Yosida theorem*, is suggested by this functional calculus perspective: since

$$(s - a)^{-1} = \int_0^{\infty} e^{-st} e^{ta} \, dt$$

for any complex number a, for s sufficiently large, we expect the generator of a strongly continuous semigroup, A, to have nonempty resolvent set, with

$$(s - A)^{-1} = \int_0^{\infty} e^{-st} e^{tA} \, dt,$$

for s sufficiently large. (See Chapter XVII.)

2

Thus the Laplace transform is useful in studying semigroups of operators (see also Chapters X and XI). We shall see that the Fourier transform and (unbounded analogues of) the Cauchy integral formula play equally significant roles (see Chapters XII, XIII, XIV, XXII, and XXIII). The idea is to use classical analysis and the intuition suggested by the functional calculus perspective to write down operator-valued formulas, and then make sense out of them. Well-known examples include the construction of holomorphic semigroups, and fractional powers using the Cauchy integral formula and constructing e^{tA^2}, when A generates a strongly continuous group, using the Fourier inversion theorem applied to e^{-y^2};

$$e^{sA^2} x = (4\pi s)^{-\frac{1}{2}} \int_{\mathbf{R}} (e^{tA}x) e^{\frac{-t^2}{4s}} \, dt,$$

for $x \in X, s > 0$.

The power series is also useful, for initial data x such that

$$\sum_{k=0}^{\infty} \frac{t^k}{k!} \|A^k x\| < \infty.$$

Then we may use (1.2) to define

$$e^{tA} x \equiv \sum_{k=0}^{\infty} \frac{t^k}{k!} A^k x.$$

(See Chapters VII, VIII and IX.)

Informally, a functional calculus for A consists of taking a complex-valued formula, and replacing the complex variable with A; sometimes a great deal of imagination is required to make sense of the resulting formula.

Thus functional calculi lead to strongly continuous semigroups, which lead to solutions of the abstract Cauchy problem. Are these implications reversible? It is well known that the existence of mild solutions of the abstract Cauchy problem, for all initial data, implies that A generates a strongly continuous semigroup (see Chapter IV for a new, greatly simplified proof of this fundamental result). What about the relationship between strongly continuous semigroups and functional calculi? The spectral theorem is an example where generating a certain type of semigroup implies the existence of a certain functional calculus. We shall see that a generalization of strongly continuous semigroups, known as *regularized semigroups*, besides their more obvious applications to the abstract Cauchy problem, may be used to construct functional calculi for unbounded operators.

3

Another approach to the abstract Cauchy problem (0.1) is the following. Suppose all solutions are unique, and let Z be the set of all initial data, x, for which the abstract Cauchy problem has a solution. In Chapter IV, we call this the *solution space* for a closed operator A. For any $t \geq 0$, define a map from Z into X, $e^{tA}x \equiv u(t,x)$. Thus, rather than consider the solutions one point at a time, we scoop them together, and consider the operator mapping a point to its corresponding solution at time t. This is the idea behind strongly continuous semigroups. When Z equals X, it can be shown that $\{e^{tA}\}_{t \geq 0}$ is a strongly continuous family contained in $B(X)$. In general, one could consider e^{tA} as an unbounded operator; however, to get all the heavy artillery of operator theory, we really want bounded operators; this is where the power of strongly continuous semigroups arises. Thus our strategy is to choose an operator, C, such that $\{e^{tA}C\}_{t \geq 0}$ is a strongly continuous family of operators in $B(X)$. This is what we call a *C-existence family* (Chapter II). When C commutes with A, so that $e^{tA}C = Ce^{tA}$, then this family of operators may be characterized by algebraic properties analogous to strongly continuous semigroups, and is called a *C-regularized semigroup*. The idea is that C "smooths" or "regularizes" whatever bad or uncontrolled behaviour e^{tA} may have; the more ill-posed the abstract Cauchy problem is, the more smoothing C must be, that is, the smaller its image must be.

Our main interest in the abstract Cauchy problem in this book is the case where it has a mild solution for all initial data in a large set not equal to the entire space. This leads to a generalization of strongly continuous semigroups, known as an *existence family*. A mild C-existence family for A corresponds to the abstract Cauchy problem having a mild solution for all initial data in the image of C. Note that an I-existence family is a strongly continuous semigroup. Consider the following example:

$$(Af)(x) \equiv xf(x), \, (x \in \mathbf{R}),$$

on some standard Banach space of functions on the real line, say $C_0(\mathbf{R})$. Clearly A does not generate a strongly continuous semigroup, for

$$(e^{tA}f)(x) \equiv e^{tx}f(x)$$

defines an unbounded operator, since e^{tx} is an unbounded function. However, there are solutions of the abstract Cauchy problem for all initial data in a dense set, namely for any initial data f such that $\lim_{|x| \to \infty} e^{tx}f(x) = 0$, for all $t \geq 0$. We shall see that these solutions are accessible through an existence family. Since $e^{-x^2}e^{tx}$ is a bounded function of x, for all $t \geq 0$,

$$(W(t)f)(x) \equiv e^{-x^2}e^{tx}f(x)$$

4

defines a mild C-existence family, where $C \equiv W(0)$.

This construction, along with the Fourier transform, shows that, if A generates a bounded strongly continuous group, then iA generates a mild C-existence family, for $C \equiv e^{-A^2}$, producing solutions of the abstract Cauchy problem, for all initial data in a dense set (see Chapter XII). Much more generally, we show that, for any matrix of polynomials of commuting generators of bounded strongly continuous groups, there exists a bounded operator C, with dense range, such that that matrix of operators generates a C-regularized semigroup (see Chapters XIII and XIV).

Just as we think of a strongly continuous semigroup generated by A as e^{tA}, and the solution of the abstract Cauchy problem (0.1) as $e^{tA}x$, we may think of a $g(A)$-existence family as $e^{tA}g(A)$, with the solution of the abstract Cauchy problem given by $e^{tA}g(A)y$, when $x = g(A)y$ is in the image of $g(A)$. When C commutes with A, as one would expect with $C = g(A)$, then $W(t) = e^{tA}C$ will satisfy $W(t)W(s) = CW(t+s)$; thus we are led to an algebraic definition, of a C-regularized semigroup.

Perhaps the simplest and most intuitive construction of a $g(A)$-regularized semigroup generated by A is with the Cauchy integral formula (or an unbounded analogue),

$$
e^{tA}g(A) \equiv \int e^{tw}g(w)(w - A)^{-1} \frac{dw}{2\pi i}, \tag{1.3}
$$

where we integrate over a cycle surrounding the spectrum of A. For example, when $-A$ generates a holomorphic strongly continuous semigroup, then this construction, with appropriate g, gives us an existence family for A. This can be applied to produce solutions, for all initial data in a dense set, of numerous ill-posed abstract Cauchy problems such as the backwards heat equation (see Chapter XII).

This also leads to a spectral intuition, that generalizes the spectral intuition of strongly continuous semigroups. When A generates a strongly continuous semigroup e^{tA}, then we expect the complex-valued function $z \mapsto e^{tz}$ to be bounded on the spectrum of A; thus the spectrum of A must be contained in a left half-plane. In fact, this is necessary but not sufficient; there is another sense (the numerical range, after an equivalent renorming) in which A must be contained in a left half-plane. To generate a $g(A)$-regularized semigroup, we expect the function $z \mapsto e^{tz}g(z)$ to be bounded on the spectrum of A. This simple spectral intuition can be used to generate sufficient conditions, involving only the location and rate of growth of the resolvent $(z - A)^{-1}$, that guarantee that A will generate C-regularized semigroups, with the corresponding information about the abstract Cauchy problem (see Chapter XXIII).

5

Another popular integral formula is the multivariable Fourier inversion formula

$$f(\vec{y}) = (2\pi)^{-\frac{n}{2}} \int_{\mathbf{R}^n} e^{i\vec{x}\cdot\vec{y}} \mathcal{F}f(\vec{x})\,dx,$$

where $\mathcal{F}$ is the Fourier transform.

If $iA_1,\ldots,iA_n$ generate commuting bounded strongly continuous groups, we define, for any polynomial p, in n variables, a $g(\vec{A})$-regularized semigroup generated by $p(A_1,\ldots,A_n)$ by

$$e^{tp(A)}g(\vec{A}) = (2\pi)^{-\frac{n}{2}} \int_{\mathbf{R}^n} e^{i\vec{x}\cdot\vec{A}} \mathcal{F}(e^{tp(\vec{y})}g(\vec{y}))(\vec{x})\,dx \qquad (1.4)$$

(see Chapters XII, XIII and XIV).

We may also use (1.2) to define $e^{tA}g(A)$, or more generally, $e^{tA}C$:

$$e^{tA}C \equiv \sum_{k=0}^{\infty} \frac{t^k}{k!} A^k C,$$

whenever

$$\sum_{k=0}^{\infty} \frac{t^k}{k!} \|A^k C\| < \infty.$$

(See Chapters VII, VIII and IX.)

Thus functional calculus constructions give us existence families. We may turn this around: we show how we may use regularized semigroups to define functional calculi for unbounded operators, via $f(A)$ defined to be the generator of $e^{tf(A)}g(A)$. That is, we define $f(A)$ by defining a regularized semigroup that it generates; use (1.3) with e^{tA} replaced by $e^{tf(A)}$ and e^{tw} replaced by $e^{tf(w)}$,

$$e^{tf(A)}g(A) \equiv \int e^{tf(w)}g(w)(w-A)^{-1}\,\frac{dw}{2\pi i},$$

or (1.4) with p replaced by f,

$$e^{tf(A)}g(\vec{A}) = (2\pi)^{-\frac{n}{2}} \int_{\mathbf{R}^n} e^{i\vec{x}\cdot\vec{A}} \mathcal{F}(e^{tf(\vec{y})}g(\vec{y}))(\vec{x})\,dx.$$

Because regularized semigroups are so much more general than strongly continuous semigroups, this enables us to define functional calculi for large classes of (unbounded) operators and large classes of functions (see Chapters XII and XXII).

6

II. EXISTENCE FAMILIES

In this chapter, we introduce a family of operators that corresponds to the abstract Cauchy problem (0.1) having a solution, for all initial data in the image of a specified bounded operator C (see Theorems 2.6 and 4.13).

Throughout this chapter, C will be a bounded operator.

First, we should make it clear what we mean by a solution of the abstract Cauchy problem.

Definition 2.1. By a *solution* of the abstract Cauchy problem (0.1) we mean a map $t \mapsto u(t,x) \in C([0,\infty),[\mathcal{D}(A)]) \bigcap C^1([0,\infty),X)$, satisfying (0.1). By a *mild solution* of (0.1) we mean a map $t \mapsto u(t,x) \in C([0,\infty),X)$ such that $v(t,x) \equiv \int_0^t u(s,x)\,ds \in \mathcal{D}(A)$, for all $t \geq 0$, and

$$\frac{d}{dt}v(t,x) = A(v(t,x)) + x, \ (t \geq 0). \tag{2.2}$$

Note that automatically $v \in C([0,\infty),[\mathcal{D}(A)])$.

We want the solutions to be well-posed, in some sense; we will discuss this more in Chapters IV and V. In stating precisely what it means to be well-posed: a small change in the initial data x should produce a small change in the solution $u(t,x)$; we are led naturally to the notion of a bounded operator; specifically, we want the map $x \mapsto u(t,x)$, which we think of as e^{tA}, to be bounded. This is despite the fact that we distinctly do *not* expect the operator A to be bounded. In fact, the existence of a unique mild solution of the abstract Cauchy problem, for all initial data, in some Frechet space, corresponds precisely to the map $T(t)x \equiv u(t,x)$ defining a strongly continuous semigroup of operators, $\{T(t)\}_{t\geq 0}$, generated by A (see Corollary 4.11).

But once we leave this idealized situation, then the map $t \mapsto u(t,x)$ is no longer a bounded operator. It may still be possible to think of e^{tA} as an *unbounded* operator (see the Notes; also see Definitions 22.6 and 7.5 and Terminology 27.1). However, to regain something analogous to well-posedness, we really need a family of bounded operators.

We wish to deal with the case where a large (possibly dense) set of initial data, that does not happen to equal the entire space, produces unique solutions or mild solutions of the abstract Cauchy problem. Suppose $Im(C)$ is contained in this set of initial data. We cannot expect the map $x \mapsto u(t,x)$ to be bounded; however, it can be shown that, if A is closed, then $x \mapsto u(t,Cx)$ is bounded (see Theorem 4.13). Thus we make the following definitions. Note that we think of $W(t)$ as being $e^{tA}C$.

7

Definition 2.3. The family of operators $\{W(t)\}_{t\geq 0} \subseteq B(X)$ is a *mild C-existence family for A* if

(1) the map $t \mapsto W(t)x$, from $[0,\infty)$ into X, is continuous, for all $x \in X$; and

(2) for all $x \in X, t > 0, \int_0^t W(s)x\,ds \in \mathcal{D}(A)$, with

$$A\left(\int_0^t W(s)x\,ds\right) = W(t)x - Cx.$$

In particular, $u(t, Cx) \equiv W(t)x$ is a mild solution of the abstract Cauchy problem, for any $x \in X$.

Definition 2.4. The family of operators $\{W(t)\}_{t\geq 0} \subseteq B([\mathcal{D}(A)])$ is a *C-existence family for A* if

(1) the map $t \mapsto W(t)x$, from $[0,\infty)$ into $[\mathcal{D}(A)]$, is continuous, for all $x \in \mathcal{D}(A)$; and

(2) for all $x \in \mathcal{D}(A), t \geq 0,$

$$\int_0^t AW(s)x\,ds = W(t)x - Cx.$$

In particular, $u(t, Cx) \equiv W(t)x$ is a solution of the abstract Cauchy problem, for any $x \in X$.

Definition 2.5. We will say that the mild C-existence family for A, $\{W(t)\}_{t\geq 0}$, is a *strong C-existence family for A* if $\{W(t)|_{[\mathcal{D}(A)]}\}_{t\geq 0}$ is a C-existence family for A.

It is obvious that these families are producing solutions of the abstract Cauchy problem (0.1), for certain initial data. We also gain uniform control over the solutions, an analogue of well-posedness: small changes in the initial data produce small changes in the solutions; however, a different yardstick is used to measure the initial data, than is used to measure the solutions (see Theorem 2.6). Clearly the goal, both in producing as many solutions as possible and in obtaining more desirable analogues of well-posedness, is to choose C whose image is as large as possible. The choice of C tells us how far from well-posedness we are.

In Theorem 4.13, we shall see that, when the solutions of the abstract Cauchy problem are unique, having a mild C-existence family is equivalent to the abstract Cauchy problem being well-posed in the strongest sense of strongly continuous semigroups, on a Frechet subspace, Z, continuously embedded in X, that contains the image of C (when the existence family

8

is exponentially bounded, Z will be a Banach space). In terms of well-posedness, this is saying that there exists a method of measurement (the topology of this subspace) with respect to which small changes in initial data produce small changes in the solution, with the same method of measurement being used. The choice of C also gives an approximation of the topology of Z, because the natural topology on the image of C, $\|y\|_{[Im(C)]} \equiv \inf\{\|x\| \mid Cx = y\}$, is stronger than the topology on Z.

Theorem 2.6.

(1) *Suppose there exists a mild C-existence family, $\{W(t)\}_{t\geq0}$, for A. Then there exists a mild solution of the abstract Cauchy problem (0.1), for all $x \in Im(C)$. The sequence of solutions $u(t, Cx_n) \equiv W(t)x_n \to 0$, uniformly on compact subsets of $[0,\infty)$, whenever $x_n \to 0$.*

(2) *Suppose there exists a C-existence family, $\{W(t)\}_{t\geq0}$, for A and A is closed. Then there exists a solution of (0.1), for all $x \in C(\mathcal{D}(A))$. Both $u(t, Cx_n) \equiv W(t)x_n$ and $Au(t, Cx_n)$ converge to 0, uniformly on compact subsets of $[0,\infty)$, whenever Ax_n and x_n converge to 0.*

See Theorem 4.13 for a converse to Theorem 2.6(a); in fact, the analogue of well-posedness will follow automatically, when A is closed and (0.1) has a mild solution, for all $x \in Im(C)$.

It is natural to want to know when a mild existence family is a strong existence family. The following proposition states a correlation between $W(t)x$ being differentiable and x being in the domain of A. Corollary 2.8 then provides sufficient conditions for a mild existence family to be a strong existence family.

Proposition 2.7. *Suppose A is closed, $\{W(t)\}_{t\geq0}$ is a mild C-existence family and the map $t \mapsto W(t)x$, from $[0,\infty)$ into X, is differentiable at $t = t_0$. Then $W(t_0)x \in \mathcal{D}(A)$ and $AW(t_0)x = \frac{d}{dt}(W(t)x)|_{t=t_0}$.*

Corollary 2.8. *Suppose A is closed, $\{W(t)\}_{t\geq0}$ is a mild C-existence family for A and either*

(1) *the map $t \mapsto W(t)x \in C^1([0,\infty), X)$, for all $x \in \mathcal{D}(A)$;*

(2) *the map $t \mapsto W(t)x \in C([0,\infty), [\mathcal{D}(A)])$, for all $x \in \mathcal{D}(A)$; or*

(3) *$W(t)A \subseteq AW(t)$, for all $t \geq 0$.*

Then $\{W(t)\}_{t\geq0}$ is a strong C-existence family for A.

A strongly continuous semigroup is defined algebraically, but it can be shown to be an I-existence family and a mild I-existence family. In the

9

next section, we will show that, when A and C commute, then an algebraic definition (*regularized semigroup*; see Definition 3.1), generalizing the algebraic definition of a strongly continuous semigroup, is equivalent to the definition of a (mild) C-existence family (see Theorems 3.5(1), 3.7 and 3.8).

In general, we are not guaranteed uniqueness of the solutions. However, when A commutes with its existence family, we are (see Theorems 3.5(2) and 3.7(a)).

If we focus on exponentially bounded solutions, then we are always guaranteed uniqueness, that is, at most one exponentially bounded solution.

Proposition 2.9. *Suppose A is closed and there exists $\omega \in \mathbf{R}$ such that A has no eigenvalues in (ω, ∞). Then all exponentially bounded solutions and mild solutions of the abstract Cauchy problem (0.1) are unique.*

Lemma 2.10. *Suppose A is closed, $u(t)$ is a $O(e^{\omega t})$ mild solution of (0.1) and $Re(z) > \omega$. Then $\int_0^\infty e^{-zt} u(t)\, dt \in \mathcal{D}(A)$ and*

$$(z - A) \int_0^\infty e^{-zt} u(t)\, dt = x.$$

Proof: Fix z such that $Re(z) > \omega$ and let $L \equiv \int_0^\infty e^{-zt} u(t)\, dt$. An integration by parts shows that L equals $z \int_0^\infty e^{-zt} \int_0^t u(s)\, ds\, dt$; thus, since A is closed, L is in $\mathcal{D}(A)$ and $AL = z \int_0^\infty e^{-zt} A(\int_0^t u(s)\, ds)\, dt = z \int_0^\infty e^{-zt}(u(t) - x)\, dt = zL - x$. ∎

Example 2.11. On $X_1 \times X_2$, define

$$A = \begin{bmatrix} G_1 & B \\ 0 & G_2 \end{bmatrix}, \qquad \mathcal{D}(A) \equiv \mathcal{D}(G_1) \times \mathcal{D}(G_2),$$

where $\mathcal{D}(G_2) \subseteq \mathcal{D}(B)$, B is a closed operator from (a subspace of) X_2 into X_1, and G_i generates a strongly-continuous semigroup, for $i = 1, 2$. Then there exists a strong C-existence family for A, for s sufficiently large, where

$$C \equiv \begin{bmatrix} I & 0 \\ 0 & (s - G_2)^{-1} \end{bmatrix}.$$

Remark 2.12. The operator A, of Example 2.11, generates a $(\lambda - A)^{-1}$-regularized semigroup (see Definition 3.1), for λ sufficiently large, and an integrated semigroup (see Chapter XVIII). Either of these facts yields unique solutions of the abstract Cauchy problem, for initial data

10

in $\mathcal{D}(A^2) = \mathcal{D}(G_1^2) \times \mathcal{D}(G_2^2)$. Example 2.11 yields unique solutions of the abstract Cauchy problem, for initial data in $C(\mathcal{D}(A)) = \mathcal{D}(G_1) \times \mathcal{D}(G_2^2)$.

Proof of Theorem 2.6:   It is obvious that $u(t, Cy) \equiv W(t)y$ is a (mild) solution of (0.1), with $x = Cy$, if $\{W(t)\}_{t\geq 0}$ is a (mild) C-existence family for A. By the uniform boundedness principle, $\|W(t)\|$ is bounded on compact subsets of $[0,\infty)$, if $\{W(t)\}_{t\geq 0}$ is a mild C-existence family, and $\|W(t)\|_{[\mathcal{D}(A)]}$ is bounded on compact subsets of $[0,\infty)$, if $\{W(t)\}_{t\geq 0}$ is a C-existence family and A is closed. This implies the analogues of well-posedness in (a) and (b). ∎

Proof of Proposition 2.7: For any $k \in \mathbf{N}$, let $x_k \equiv k \int_{t_0}^{t_0+\frac{1}{k}} W(s)x \, ds$. Then x_k converges to $W(t_0)x$, while $Ax_k = k\left(W(t_0 + \frac{1}{k})x - W(t_0)x\right)$ converges to $\frac{d}{dt}W(t)x|_{t=t_0}$, as $k \to \infty$. Thus this result follows from the fact that A is closed. ∎

Proof of Corollary 2.8: For (1), this follows from Proposition 2.7, since we may then integrate $AW(s)x = \frac{d}{ds}W(s)x$, from 0 to t. For (2), this follows from the fact that A is closed and $s \mapsto AW(s)x$, from $[0,\infty) \to X$, is continuous, so that $A(\int_0^t W(s)x \, ds) = \int_0^t AW(s)x \, ds$, for $x \in \mathcal{D}(A)$. Note finally that hypothesis (3) implies hypothesis (2). ∎

Proof of Proposition 2.9: Suppose u is an exponentially bounded solution or mild solution of (0.1), with $u(0) = 0$. Then by Lemma 2.10, its Laplace transform is trivial, for $Re(z)$ sufficiently large, which implies that $u \equiv 0$, as desired. ∎

Proof of Example 2.11:   Formally,

$$e^{tA} = \begin{bmatrix} e^{tG_1} & \int_0^t e^{(t-w)G_1} B e^{wG_2} \, dw \\ 0 & e^{tG_2} \end{bmatrix}.$$

To make this bounded, we need an $(s - G_2)^{-1}$ to the right of B. This explains our choice of C.

For $t \geq 0$, let

$$W(t) \equiv \begin{bmatrix} e^{tG_1} & \int_0^t e^{(t-w)G_1} B(s-G_2)^{-1} e^{wG_2} \, dw \\ 0 & e^{tG_2}(s-G_2)^{-1} \end{bmatrix}.$$

For any continuously differentiable $f : [0,\infty) \mapsto X_i$ ($i = 1, 2$), $t \geq 0$, it is well known that $\int_0^t e^{rG_i} f(r) \, dr \in \mathcal{D}(G_i)$, with

$$G_i \left(\int_0^t e^{rG_i} f(r) \, dr \right) = e^{tG_i} f(t) - f(0) - \int_0^t e^{rG_i} f'(r) \, dr \qquad (2.12)$$

11

(see Theorem 3.4(c)). This implies that, for any $x \in X_1 \times X_2, t \geq 0$, $\int_0^t W(s)x\,ds \in \mathcal{D}(A)$, and a calculation, using (2.12), implies that $W(t)$ satisfies Definition 2.3.

Thus $\{W(t)\}_{t \geq 0}$ is a mild C-existence family for A.

Also by (2.12), if $x \in \mathcal{D}(A)$, and $t \geq 0$, then $W(t)x \in \mathcal{D}(A)$, and a calculation shows that

$$AW(t)x$$

$$= \begin{bmatrix} e^{tG_1}(G_1 x_1 + B(s - G_2)^{-n}x_2) + \int_0^t e^{(t-w)G_1} B(s - G_2)^{-1} e^{wG_2} G_2 x_2\,dw \\ (s - G_2)^{-1}(e^{tG_2}G_2 x_2) \end{bmatrix}.$$

This is a continuous function of t. Thus $t \mapsto W(t)x$ is a continuous map from $[0, \infty)$ into $[\mathcal{D}(A)]$, for all x in $\mathcal{D}(A)$. It is straightforward to show that A is closed. Thus, by Corollary 2.8(2), $\{W(t)\}_{t \geq 0}$ is a strong C-existence family for A. ∎

III. REGULARIZED SEMIGROUPS

Throughout this chapter, C will be a bounded, injective operator. We think of a C-existence family as being $W(t) = e^{tA}C$. When $e^{tA}C = Ce^{tA}$, a very short calculation shows (at least if A is a complex number) that $CW(t + s) = W(t)W(s)$. Thus we are led to the following algebraic definition, which is obviously analogous to the definition of a strongly continuous semigroup (an I-regularized semigroup).

Definition 3.1. The strongly continuous family $\{W(t)\}_{t \geq 0} \subseteq B(X)$ is a *C-regularized semigroup* if

(1) $W(0) = C$, and

(2) $W(t)W(s) = CW(t + s)$, for all $s, t \geq 0$.

A *generates* $\{W(t)\}_{t \geq 0}$ if

$$Ax = C^{-1}\left[\lim_{t \to 0} \frac{1}{t}\left(W(t)x - Cx\right)\right],$$

with

$$\mathcal{D}(A) = \{x|\ \text{the limit exists and is in } \mathrm{Im}(C)\}.$$

When we do not wish to specify the C, we will say merely *regularized semigroup*.

We will show that, if A generates a strongly continuous semigroup $\{e^{tA}\}_{t \geq 0}$ that commutes with C, then A is the generator of the C-regularized semigroup $\{Ce^{tA}\}_{t \geq 0}$. More generally, if A generates a C_1-regularized semigroup $\{W(t)\}_{t \geq 0}$ that commutes with C_2, then A generates the $C_1 C_2$-regularized semigroup $\{C_2 W(t)\}_{t \geq 0}$ (Proposition 3.10).

Example 3.2. The following simple example demonstrates how broad a class of operators and abstract Cauchy problems may be dealt with using regularized semigroups.

Define $Af(x) \equiv xf(x)$, on $C_0(\mathbf{R})$. Then

$$W(t)f(x) \equiv e^{-x^2}e^{tx}f(x)$$

is a $W(0)$-regularized semigroup generated by A.

A similar construction, aided by the Fourier transform or the Cauchy integral formula, shows that, if iA generates a strongly continuous group, then A generates an e^{-A^2}-regularized semigroup (see Chapter XIV and Example 22.24).

13

Example and Remarks 3.3. Example 3.2 addresses the first objection to regularized semigroups: "Does anything generate a regularized semigroup that does not generate a strongly continuous (or integrated; see Chapter XVIII) semigroup?" The subsequent objection is usually: "Everything generates a regularized semigroup, therefore it's a meaningless concept; to say that something generates a regularized semigroup says nothing." Here is a simple example of an operator that does not generate a regularized semigroup.

Define $Af \equiv f'$, on $X \equiv \{f \in C[0,1] | f(0) = 0\}$, with maximal domain. Then it is not hard to see that the abstract Cauchy problem (0.1) has no nontrivial solutions; thus by Theorem 3.5, A cannot generate a C-regularized semigroup, for any C.

So not every closed operator generates a regularized semigroup. *If the abstract Cauchy problem has mild solutions, for all initial data in the image of C, and C commutes with A, then an extension of A generates a C-regularized semigroup* (see Theorem 4.15). Thus information about generating a regularized semigroup is the most fundamental information, telling us whether or not to expect nontrivial solutions of the abstract Cauchy problem. Much more than this, the *choice* of C tells us how many solutions to expect (see Theorems 3.5, 3.13 and 3.14); the larger the image of C, the more solutions we are guaranteed. Thus, if we say, instead of, "A generates a C-regularized semigroup, for some C," rather, "A generates a C-regularized semigroup, for this specific choice of C," then we have gained very specific and worthwhile information about A and the corresponding abstract Cauchy problem.

Rather than merely saying, for a given A, that the abstract Cauchy problem is or is not well posed (does or does not generate a strongly continuous semigroup), it is more constructive to identify C such that A generates a C-regularized semigroup; the choice of C provides a continuous scale of measurements of how far from being well-posed the abstract Cauchy problem is.

As we shall see later (Chapter IV), a C-regularized semigroup generated by A may be used to construct a Frechet space (a Banach space if the regularized semigroup is exponentially bounded), Z, continuously embedded in X, on which A generates a strongly continuous semigroup, so that the usual form of well-posedness occurs. Again the choice of C is informative, because the image of C will be contained in Z.

Theorem 3.4. *Suppose $\{W(t)\}_{t \geq 0}$ is a C-regularized semigroup generated by A. Then*

(a) *A is closed;*

(b) *$Im(C) \subseteq \overline{\mathcal{D}(A)}$;*

14

(c) if $f : [0, \infty) \to X$ is continuously differentiable and $t \geq 0$, then $\int_0^t W(s)f(s)\,ds \in \mathcal{D}(A)$, with

$$A\left[\int_0^t W(s)f(s)\,ds\right] = W(t)f(t) - Cf(0) - \int_0^t W(s)f'(s)\,ds;$$

and

(d) for all $t \geq 0$, $W(t)A \subseteq AW(t)$, with

$$W(t)x = Cx + \int_0^t W(s)Ax\,ds,$$

for all $x \in \mathcal{D}(A)$.

It is clear from Theorem 3.4(c) and (d) that $\{W(t)\}_{t\geq 0}$ is a strong C-existence family for A. Thus Theorem 2.6 guarantees the existence of solutions and mild solutions of the abstract Cauchy problem. With regularized semigroups, we are also guaranteed uniqueness of the solutions, even if merely an extension of A generates a C-regularized semigroup.

Theorem 3.5.

(1) If $\{W(t)\}_{t\geq 0}$ is a C-regularized semigroup generated by A, then $\{W(t)\}_{t\geq 0}$ is a strong C-existence family for A.

(2) If an extension of A generates a regularized semigroup, then all solutions and mild solutions of the abstract Cauchy problem (0.1) are unique.

In Theorem 3.7 we give a partial converse to 3.4(c) and (d). First we must introduce the C-resolvent set, which will play the same role with C-regularized semigroups that the resolvent set plays with strongly continuous semigroups (see Chapter XVII).

Definition 3.6. The complex number λ is in $\rho_C(A)$, the C-resolvent set of A, if $(\lambda - A)$ is injective and $Im(C) \subseteq Im(\lambda - A)$.

It will often be the case that a regularized semigroup will be generated by a proper extension of the operator of interest. In the next two theorems, we investigate the relationship between having an existence family and having an extension that generates a regularized semigroup.

In (b) of the following, note that, whenever $\{W(t)\}_{t\geq 0}$ is an exponentially bounded C-regularized semigroup, then $\rho_C(A)$ is nonempty (Proposition 17.2).

15

Theorem 3.7. *Suppose A is closed and $\{W(t)\}_{t\geq 0}$ is a strongly continuous family of bounded operators, with $W(t)A \subseteq AW(t)$, for all $t \geq 0$.*

(a) *If $\{W(t)\}_{t\geq 0}$ is a mild C-existence family for A, then $\{W(t)\}_{t\geq 0}$ is
a C-regularized semigroup generated by an extension of A.*

(b) *If $\{W(t)\}_{t\geq 0}$ is a C-existence family for A, then $\{W(t)\}_{t\geq 0}$ is a
C-regularized semigroup generated by an extension of A, if either*

    (1) $\mathcal{D}(A)$ is dense; or

    (2) $\rho_C(A)$ is nonempty.

Theorem 3.8. *Suppose $\{W(t)\}_{t\geq 0}$ is a C-regularized semigroup, generated by an extension of A. Then*

(1) *if $\int_0^t W(s)x\,ds \in \mathcal{D}(A)$, for all $t \geq 0, x \in X$, then $\{W(t)\}_{t\geq 0}$ is a
mild C-existence family for A;*

(2) *if $W(t)$ leaves $\mathcal{D}(A)$ invariant, for all $t \geq 0$, then $\{W(t)\}_{t\geq 0}$ is a
C-existence family for A; and*

(3) *if, in addition to (2), A is closed and densely defined, then $\{W(t)\}_{t\geq 0}$
is a strong C-existence family for A.*

The solution or mild solution is given by $u(t, Cx) = W(t)x$.

For the abstract Cauchy problem, Theorems 3.8 and 2.6 indicate how
the distinction between an extension of A being the generator and A itself
being the generator does not seem crucial; this is why the concept of an
existence family (Chapter II) is in many ways more natural for dealing
with the abstract Cauchy problem.

However, it is of interest to know when A itself is the generator. For
example, in future chapters we will define functional calculi, $f \mapsto f(A)$,
for A, by constructing, for each function f, a regularized semigroup and
then defining $f(A)$ to be the generator.

Note that the hypotheses of Proposition 3.9 are satisfied when $\rho(A)$ is
nonempty and $W(t)A \subseteq AW(t)$, for all $t \geq 0$.

Proposition 3.9. *Suppose an extension of A generates the regularized
semigroup $\{W(t)\}_{t\geq 0}$ and there exists bounded, injective G such that
$\mathcal{D}(A) = Im(G), \mathcal{D}(A^2) = Im(G^2)$, and $GW(t) = W(t)G$, for all $t \geq 0$.
Then A is the generator.*

The following proposition shows that there is no harm in applying additional smoothing, that is, the generator remains the same.

Proposition 3.10. *Suppose $\{W(t)\}_{t\geq 0}$ is a C_1-regularized semigroup
generated by A and $C_2W(t) = W(t)C_2$, for all $t \geq 0$. Then $\{C_2W(t)\}_{t\geq 0}$
is a C_1C_2-regularized semigroup generated by A.*

Proposition 3.11. *Suppose an extension of A, $\tilde{A}$, generates a C-regularized semigroup. Then the following are equivalent.*

(1) $C\left(\mathcal{D}(\tilde{A})\right) \subseteq \mathcal{D}(A)$.

(2) $C^{-1}AC = \tilde{A}$.

The hypothesis on $\mathcal{D}(A)$ in the preceding proposition does not seem like a natural one, since it requires extensive knowledge about the generator, $\tilde{A}$. In practice, all that we may have information about may be A itself. The following Corollary provides a sufficient condition that is easier to verify; in particular, if the regularized semigroup is exponentially bounded, then $\rho_C(\tilde{A})$ automatically contains a left half-plane (see Proposition 17.2).

Corollary 3.12. *Suppose A is closed, $A \subseteq \tilde{A}$, $\tilde{A}$ generates a C-regularized semigroup and $\rho_C(A) \cap \rho_C(\tilde{A})$ is nonempty. Then $\tilde{A} = C^{-1}AC$.*

It is interesting that we may use a regularized semigroup generated by A to characterize *all* solutions of the abstract Cauchy problem.

We also obtain uniform control over the "smoothed" or regularized solutions Cu, for *all* initial data x for which the abstract Cauchy problem has a solution. See Chapters IV and V for *bona fide* well-posedness of the solutions, at least on a subspace, when A generates a regularized semigroup.

In the following two theorems, *any* C such that A generates a C-regularized semigroup may be used. This is useful, because, if one allows the image of C to be small, it is easier to obtain a C-regularized semigroup; but regardless of how restrictive C is, we still may use the C-regularized semigroup to obtain all solutions of the abstract Cauchy problem; that is, it is not necessary to try to choose an optimal C, in any sense.

Theorem 3.13. *Suppose A generates a C-regularized semigroup $\{W(t)\}_{t\geq 0}$. Then the following are equivalent:*

(1) *there exists a mild solution of the abstract Cauchy problem (0.1);*

(2) $W(t)x \in Im(C)$, *for all $t \geq 0$, and $t \mapsto C^{-1}W(t)x$, from $[0,\infty)$ into X, is continuous.*

The solution is given by

$$u(t,x) = C^{-1}W(t)x.$$

Whenever $\{x_n\}_{n=1}^{\infty}$ is a sequence of initial data, for which (0.1) has a solution, converging to zero, then $Cu(t,x_n)$ converges to zero, uniformly on compact subsets of $[0,\infty)$.

17

Theorem 3.14. *Suppose A generates a C-regularized semigroup $\{W(t)\}_{t \geq 0}$. Then the following are equivalent:*

(1) *there exists a solution of the abstract Cauchy problem (0.1);*

(2) *$W(t)x \in C(\mathcal{D}(A))$, for all $t \geq 0$ and $t \to C^{-1}AW(t)x$, from $[0, \infty)$ into X, is continuous; and*

(3) *$W(t)x \in Im(C)$, for all $t \geq 0$ and $t \mapsto C^{-1}W(t)x$, from $[0, \infty)$ into X, is differentiable.*

The solution is given by

$$u(t, x) = C^{-1}W(t)x.$$

Whenever $\{x_n\}$ is a sequence of initial data, for which (0.1) has a solution, converging to zero, then $Cu(t, x_n)$ converges to zero, uniformly on compact subsets of $[0, \infty)$.

Proof of Theorem 3.4: (d) Fix $t \geq 0, x \in \mathcal{D}(A)$. For $s \geq 0$,

$$\frac{1}{s}[W(s)(W(t)x) - C(W(t)x)] = W(t)\left[\frac{1}{s}(W(s)x - Cx)\right],$$

so that, since $W(t)$ is bounded,

$$\lim_{s \to 0} \frac{1}{s}[W(s)(W(t)x) - C(W(t)x)]$$

exists and equals $W(t)CAx = CW(t)Ax$. This implies that $W(t)x \in \mathcal{D}(A)$, with $AW(t)x = W(t)Ax$.

Note also that

$$\frac{1}{s}[CW(s+t)x - CW(t)x] = \frac{1}{s}[W(s)W(t)x - CW(t)x],$$

so that $\frac{d}{ds}CW(s)x|_{s=t}$ exists, and equals $CW(t)Ax$. Thus, for $t \geq 0, x \in \mathcal{D}(A)$,

$$CW(t)x = C^2x + \int_0^t CW(s)Ax\,ds = C\int_0^t W(s)Ax\,ds,$$

since C is bounded. Since C is injective, this implies (d).

(a) Suppose $\{x_n\}_{n=1}^{\infty} \subseteq \mathcal{D}(A)$, with $x_n \to x$ and $Ax_n \to y$. Fix $t > 0$. By the uniform boundedness principle, $\|W(s)\|$ is bounded on $[0, t]$. Thus, $\{W(s)x_n\}$ converges to $W(s)x$, and $\{\frac{d}{ds}W(s)x_n\} = \{W(s)Ax_n\}$ (by (d))

18

converges to $W(s)y$, as $n \to \infty$, both uniformly on $[0,t]$. This implies that

$$W(t)x - Cx = \lim_{n \to \infty} (W(t)x_n - Cx_n)$$

$$= \lim_{n \to \infty} \int_0^t W(s)Ax_n \, ds$$

$$= \int_0^t W(s)y \, ds,$$

so that $\lim_{t \to 0} \frac{1}{t}(W(t)x - Cx)$ exists and equals Cy. This implies that $x \in \mathcal{D}(A)$, with $Ax = y$, as desired.

(c) For fixed t, let $x \equiv \int_0^t W(s)f(s) \, ds$. For $h > 0$,

$$\frac{1}{h}(W(h)x - Cx) = \frac{1}{h}\left[\int_0^t (CW(s+h) - CW(s))f(s)\,ds\right]$$

$$= \frac{1}{h}\left[\int_h^{t+h} CW(w)f(w-h)\,dw - \int_0^t CW(s)f(s)\,ds\right]$$

$$= \frac{1}{h}\int_h^t CW(s)(f(s-h) - f(s))\,ds$$

$$+ \frac{1}{h}\int_t^{t+h} W(w)f(w-h)\,dw - \frac{1}{h}\int_0^h CW(s)f(s)\,ds,$$

so that $\lim_{h \to 0}$ exists, with

$$\lim_{h \to 0} \frac{1}{h}(W(h)x - Cx) = \int_0^t -CW(s)f'(s)\,ds + CW(t)f(t) - C^2 f(0).$$

Thus $x \in \mathcal{D}(A)$, with $Ax = -\int_0^t W(s)f'(s)\,ds + W(t)f(t) - Cf(0)$. ∎

(b) For all $x \in X$, $\lim_{t \to 0} \frac{1}{t}\int_0^t W(s)x \, ds = W(0)x = Cx$. Thus, (b) follows from (c). ∎

Proof of Theorem 3.5: (a) is clear from Theorem 3.4. For (b), suppose $\tilde{A}$ is the generator of the C-regularized semigroup $\{W(t)\}_{t \geq 0}$ and $A \subseteq \tilde{A}$. Suppose $v'(t) = A(v(t))$, for all $t \geq 0$ and $v(0) = 0$. For $0 \leq s \leq t$,

$$\frac{d}{ds}W(t-s)v(s) = W(t-s)A(v(s)) - \tilde{A}W(t-s)v(s) = 0,$$

19

by Theorem 3.4. Thus $Cv(t) = W(t)v(0) = 0$, so that, since C is injective, $v(t) \equiv 0$, proving uniqueness. ∎

Proof of Theorem 3.7: (a) Since A is closed and $W(t)A \subseteq AW(t)$, by Corollary 2.8, $\{W(t)\}_{t \geq 0}$ is a strong C-existence family for A, with

$$W(t)x = Cx + \int_0^t W(s)Ax\,ds, \tag{3.15}$$

for all $t \geq 0, x \in \mathcal{D}(A)$, as in (b). To see that $\{W(t)\}_{t \geq 0}$ is a C-regularized semigroup, suppose $s, t \geq 0$ and $x \in X$. We calculate as follows:

$$\int_s^{s+t} W(r)Cx\,dr = \int_0^t W(s+t-w)Cx\,dw$$

$$= \int_0^t \frac{d}{dw}\left[W(s+t-w)\left(\int_0^w W(r)x\,dr\right)\right]dw$$

$$= W(s)\left(\int_0^t W(r)x\,dr\right),$$

so that differentiating with respect to s implies that

$$W(s+t)Cx - W(s)Cx = W(s)A\left(\int_0^t W(r)x\,dr\right) = W(s)\left(W(t)x - Cx\right).$$

Thus $\{W(t)\}_{t \geq 0}$ is a C-regularized semigroup. It is clear from (3.15) that its generator is an extension of A.

(b) As in (a),

$$W(t)W(s)x = CW(t+s)x, \ W(0)x = Cx, \ \forall x \in \mathcal{D}(A), \ s, t \geq 0.$$

If $\mathcal{D}(A)$ is dense, then the same is true for all $x \in X$, since $W(t)$ and C are bounded. If there exists $\lambda \in \rho_C(A)$, then, since $W(t)A \subseteq AW(t)$, $W(t)$ commutes with $(\lambda - A)^{-1}C$; thus, for any $x \in X$,

$$(\lambda - A)^{-1}C\left[W(t)W(s)x\right] = W(t)W(s)(\lambda - A)^{-1}Cx$$

$$= CW(t+s)(\lambda - A)^{-1}Cx$$

$$= (\lambda - A)^{-1}C\left[CW(t+s)x\right],$$

so, since $(\lambda - A)^{-1}C$ is injective, $\{W(t)\}_{t \geq 0}$ satisfies (2) of Definition 3.1. An identical argument shows that $W(0) = C$. Assertion (3.15) implies that the generator is an extension of A. ∎

20

Proof of Theorem 3.8: Assertions (1) and (2) follow immediately from Theorem 3.4. For (3), all that needs to be shown is that

$$\int_0^t W(s)x\, ds \in \mathcal{D}(A), \qquad (3.16)$$

for all $x \in X, t \geq 0$. Since A is closed and $W(t)A \subseteq AW(t)$, (3.16) is true for $x \in \mathcal{D}(A), t \geq 0$. For arbitrary $x \in X, t \geq 0$, let $< x_n >_{n=0}^\infty$ be a sequence in $\mathcal{D}(A)$ converging to x. Then $\int_0^t W(s)x_n\, ds$ converges to $\int_0^t W(s)x\, ds$, while

$$A\left(\int_0^t W(s)x_n\, ds\right) = W(t)x_n - Cx_n$$

(Theorem 3.4) converges. Thus, since A is closed, (3.16) follows. ∎

Proof of Proposition 3.9: Let $\tilde{A}$ be the generator of $\{W(t)\}_{t\geq 0}$. Suppose $x \in \mathcal{D}(\tilde{A})$. Then

$$CG\tilde{A}x = GC\tilde{A}x = G\left(\frac{d}{dt}W(t)x|_{t=0}\right) = \frac{d}{dt}W(t)Gx|_{t=0} = C\tilde{A}Gx$$

$$= CAGx.$$

Since C is injective, $G\tilde{A}x = AGx$, thus $AGx \in \mathcal{D}(A)$, so that $Gx \in \mathcal{D}(A^2)$. This means there exists $y \in X$ such that $Gx = G^2 y$, so that, since G is injective, $x = Gy \in \mathcal{D}(A)$, as desired. ∎

Proof of Proposition 3.10: It is clear that $\{C_2 W(t)\}_{t\geq 0}$ is a $C_1 C_2$-regularized semigroup. Let A_1 be its generator. Since C_2 is bounded, it is easy to show that $A \subseteq A_1$. Suppose $x \in \mathcal{D}(A_1)$. Then, for all $t \geq 0$, by Theorem 3.4(d),

$$C_2\left(W(t)x - C_1 x\right) = \int_0^t C_2 W(s)A_1 x\, ds$$

$$= C_2\left(\int_0^t W(s)A_1 x\, ds\right);$$

since C_2 is injective, this implies that

$$(W(t)x - Cx) = \int_0^t W(s)A_1 x\, ds,$$

which implies that $x \in \mathcal{D}(A)$, with $Ax = A_1 x$, as desired. ∎

21

Proof of Proposition 3.11: $(1) \to (2)$. Note that by definition of the generator, $C^{-1}\tilde{A}C$ is the generator of the C^2-regularized semigroup $\{CW(t)\}_{t\geq 0}$; thus, by Proposition 3.10,

$$C^{-1}\tilde{A}C = \tilde{A}. \tag{3.17}$$

Let $\mathcal{D}(B) \equiv C\left(\mathcal{D}(\tilde{A})\right), Bx = Ax$, for all $x \in \mathcal{D}(B)$. Suppose $x \in \mathcal{D}(\tilde{A})$. Then $Cx \in \mathcal{D}(B)$ and $B(Cx) = \tilde{A}Cx = C\tilde{A}x$, by Theorem 3.4(d), thus $x \in \mathcal{D}(C^{-1}BC)$ and $C^{-1}BCx = \tilde{A}x$. This is saying that

$$\tilde{A} \subseteq C^{-1}BC \subseteq C^{-1}AC \subseteq C^{-1}\tilde{A}C = \tilde{A},$$

by (3.17). This proves (2).

$(2) \to (1)$ is obvious from the definition of $\mathcal{D}(C^{-1}AC)$. ∎

Proof of Corollary 3.12: By Proposition 3.11, it is sufficient to show that $C\left(\mathcal{D}(\tilde{A})\right) \subseteq \mathcal{D}(A)$. There exists $r \in \rho_C(A) \cap \rho_C(\tilde{A})$. If $x \in \mathcal{D}(\tilde{A})$, then

$$Cx = C(r-\tilde{A})^{-1}(r-\tilde{A})x = (r-\tilde{A})^{-1}C(r-\tilde{A})x = (r-A)^{-1}C(r-\tilde{A})x,$$

which is in $\mathcal{D}(A)$. ∎

Proof of Theorem 3.13: Suppose (0.1) has a mild solution, u. Let $v(t) \equiv \int_0^t u(s)\,ds$. Then, since C is bounded and $CA \subseteq AC$,

$$\frac{d}{dt}(Cv)(t) = A(Cv(t)) + Cx \; \forall t > 0.$$

By Theorem 3.5(2),

$$Cu(t) = C\frac{d}{dt}v(t) = \frac{d}{dt}(Cv)(t) = W(t)x,$$

which implies that $t \mapsto C^{-1}W(t)x$ is continuous and $u(t) = C^{-1}W(t)x$.

Conversely, suppose $W(t)x \in Im(C)$, for all $t \geq 0$, and $t \mapsto C^{-1}W(t)x$ is continuous. Let $v(t) \equiv \int_0^t C^{-1}W(s)x\,ds$. Then by Theorem 3.4, $Cv(t) \in \mathcal{D}(A)$, with $ACv(t) = W(t)x - Cx$; thus $v(t) \in \mathcal{D}(C^{-1}AC)$, with

$$C^{-1}ACv(t) = C^{-1}W(t)x - x = v'(t) - x.$$

Since, by Proposition 3.11, $A = C^{-1}AC$, this concludes the proof. ∎

22

Proof of Theorem 3.14: Suppose (0.1) has a solution, u. Then by Theorem 3.5(2), $Cu(t) = W(t)x$, so that $W(t)x \in C(\mathcal{D}(A))$, for all $t \geq 0$, with

$$\frac{d}{dt}u(t) = C^{-1}AW(t)x,$$

so that $t \mapsto C^{-1}AW(t)x$ is continuous and $t \mapsto C^{-1}W(t)x = u(t)$ is differentiable.

Conversely, if $t \mapsto C^{-1}AW(t)x$ is defined and continuous, let $u(t) \equiv C^{-1}W(t)x$. As in the proof of Theorem 3.13, $u(t) \in \mathcal{D}(A) = \mathcal{D}(C^{-1}AC)$, with $Au(t) = C^{-1}AW(t)x$, so that, since C^{-1} is closed,

$$\int_0^t Au(s)\,ds = C^{-1}\int_0^t AW(s)x\,ds = C^{-1}(W(t)x - Cx) = u(t) - x,$$

thus u is a solution of (0.1).

If $t \mapsto C^{-1}W(t)x$ is differentiable, then by Proposition 2.7, $Cu(t) \in \mathcal{D}(A)$, with $ACu(t) = Cu'(t)$; thus $u(t) \in \mathcal{D}(C^{-1}AC) = \mathcal{D}(A)$, with

$$Au(t) = C^{-1}ACu(t) = u'(t);$$

thus again u is a solution of (0.1). ∎

23

IV. THE SOLUTION SPACE OF AN OPERATOR AND AUTOMATIC WELL-POSEDNESS

Many physical problems may be modeled as an abstract Cauchy problem (0.1), where A is an operator on a locally convex space. It is well known that, in order for this model to have any practical value, it is not enough to have plenty of solutions; (0.1) should also be *well-posed*; informally, this means that small changes in x, the initial data (corresponding to small errors in measurement) should yield small changes in $u(t, x)$.

Although there is sometimes uncertainty about what precisely constitutes well-posedness, one condition that most would agree implies well-posedness is having A generate a strongly continuous semigroup. This implies that a sequence of solutions $u(t, x_n)$ converges to $u(t, x)$, uniformly on compact subsets of $[0, \infty)$, whenever the initial data x_n converge to x.

It is often the case that A does not generate a strongly continuous semigroup. Consider the following list, which includes what are perhaps the most well-known partial differential equations: heat equation, Schrödinger equation, wave equation, Cauchy problem for the Laplace equation, backwards heat equation, all on $L^p(\Omega)$, for appropriate $\Omega \subseteq \mathbf{R}^n$, with appropriate boundary conditions. For each equation, when written as an abstract Cauchy problem (0.1), let us ask whether the operator A that appears generates a strongly continuous semigroup, for $1 \leq p < \infty$. The answers are, respectively, "yes," "sometimes," "sometimes," "no" and "no." Yet in all these cases, a unique solution exists, for all initial data in a dense set.

The class of operators that generate C-regularized semigroups is much larger than the class of operators that generate strongly continuous semigroups. In all the examples in the previous paragraph, we shall see in later chapters that A generates a C-regularized semigroup. The choice of C measures how ill-posed the problem is. When C is chosen to be $(\lambda - A)^{-n}$, for some $n \in \mathbf{N}$, then $i\Delta$, on $L^p(\mathbf{R}^k)$, for $1 \leq p \leq \infty, p \neq 2$, may be shown to generate a C-regularized semigroup, but not a strongly continuous semigroup (see Chapter XI). This yields solutions of the Schrödinger equation, for all initial data in the domain of Δ^{n+1}. Much "worse" operators, corresponding to what are traditionally referred to as ill posed or improperly posed problems, generate C-regularized semigroups. For example, if $A \equiv -\Delta$, so that (0.1) becomes the backwards heat equation, then A generates a C-regularized semigroup (see Chapter VIII). If $A \equiv \begin{bmatrix} 0 & I \\ -\Delta & 0 \end{bmatrix}$, so that (0.1) becomes the Cauchy problem for the Laplace equation, then A generates a C-regularized semigroup (see Chapter IX).

24

When $Im(C)$ is dense, as is the case in the examples above, then (0.1) has a unique solution for all x in a dense set. Thus, C-regularized semigroups may often be used to produce unique solutions for all initial data in a dense set. In this chapter and the next, we will address the question of whether the solutions are well-posed, in some sense.

In this chapter, we show that, if A is closed and there exists a nonempty set of initial data for which the abstract Cauchy problem (0.1) has a unique mild solution, then there exists a Frechet space, Z, that contains all such initial data, on which A generates a strongly continuous semigroup. This is saying that we can always *make* (0.1) well-posed, regardless of how ill-posed our original formulation was. And if one adopts the working hypothesis that all physically correct models are well-posed, this construction may be considered a way of automatically correcting one's first guess; the space on which A acts is often chosen out of convenience.

As a corollary, we obtain a much shorter, easier proof, than currently exists, of the well-known relationship between the abstract Cauchy problem and strongly continuous semigroups (Corollaries 4.11 and 4.12). This is a fundamental result that has seen many proofs, all of them somewhat involved, even when X is a Banach space.

More generally, we use the solution space to show that, for any bounded operator C, the existence of a unique solution of (0.1), for all initial data in the image of C, corresponds to A having a mild C-existence family (see Chapter II). When A and C commute and C is injective, this is a C-regularized semigroup (see Chapter III.)

The only objection to the solution space is that it is not practical, in general, to try to construct it explicitly. C-existence families and C-regularized semigroups provide a simple method of approximating the solution space and its topology, as follows.

We will write $Y \hookrightarrow X$ to mean that Y is continuously embedded in X, that is, $Y \subseteq X$ and the identity map from Y to X is continuous.

We show that, when A generates a C-regularized semigroup, then

$$[Im(C)] \hookrightarrow Z \hookrightarrow X,$$

and $A|_Z$, the restriction of A to Z, generates a strongly continuous semigroup. If the regularized semigroup is exponentially bounded, then we shall see in the next chapter that we may choose a Banach space; in general, Z is a Frechet space. If $\rho(A)$ is nonempty, the converse is also true. The norm on the solution space Z can then be expressed in terms of the regularized semigroup: $\|x\|_Z \equiv \sup\{\|C^{-1}W(t)x\| \,|\, t \geq 0\}$, where $\{W(t)\}_{t \geq 0}$ is the C-regularized semigroup generated by A.

This chapter shows that, in a technical sense, at least if one is willing to make renormings, the concepts of existence family and strongly continuous

25

semigroup are the same. But there are important practical differences. An existence family is often very easy to produce and construct, and one does not have to leave the original norm, which may be very simple or physically meaningful. The solution space, on which the restriction of A generates a strongly continuous semigroup, may be very difficult to construct, with a norm that is unpleasant or impossible to work with. Throughout this book, there are numerous examples of regularized semigroups that are constructed in the most simple-minded and intuitive manner.

Even if one's only goal is to find subspaces, Y, on which the restriction of A, $A|_Y$, generates a strongly continuous semigroup, we submit the following algorithm. First find a regularized semigroup generated by A. Then use the construction of this chapter (Proposition 4.14) to produce Y.

This chapter makes it clear how regularized semigroups may be used, first, to characterize all initial data that yield a solution of the abstract Cauchy problem (0.1) (see Proposition 4.14), then, to find a norm with respect to which those solutions are well-posed. Perhaps more importantly, for practical purposes, is the fact that the choice of C, when A generates a C-regularized semigroup, tells one how to measure how far from well-posedness one is, and to approximate the space on which A is well-posed.

In this chapter only, the space X will always be a Frechet space, topologized by the seminorms $\{\|\ \|_j\}_{j \in \mathbb{N}}$. We shall see that this is the natural generality for these results (see Remark 4.9 and Example 4.10). In the next chapter, we will discuss how to obtain well-posedness on a continuously embedded Banach subspace, when the original space is a Banach space.

Throughout this chapter, A will always be a closed operator, $C \in B(X)$.

Definition 4.1. We will write $Z \hookrightarrow X$ to mean that $Z \subseteq X$ and Z is continuously embedded in X, that is, the identity map, from Z into X, is continuous.

Definition 4.2. By $A|_Z$ we will mean the part of A in Z, that is, $\mathcal{D}(A|_Z)$ equals $\{x \in \mathcal{D}(A) \cap Z \mid Ax \in Z\}$ and $A|_Z x = Ax$, for all $x \in \mathcal{D}(A|Z)$.

Definition 4.3. We will write $[\mathcal{D}(A)]$ for the Frechet space $\mathcal{D}(A)$ with the graph topology, that is, topologized by the seminorms

$$\|x\|_{[\mathcal{D}(A)],j} \equiv \|x\|_j + \|Ax\|_j.$$

We will write $[Im(C)]$ for the Frechet space topologized by the seminorms

$$\|x\|_{[Im(C)],j} \equiv \inf\{\|y\|_j \mid Cy = x\}.$$

26

Definition 4.4. By $C([0,\infty), X)$ we will mean the Frechet space topologized by the seminorms

$$\|\phi\|_{a,b,j} \equiv \sup_{t\in[a,b]} \|\phi(t)\|_j, \ a,b \in \mathbf{Q}^+, j \in \mathbf{N}.$$

Definition 4.5. The strongly continuous family of bounded operators $\{T(t)\}_{t\geq 0}$, on a locally convex space Y, is a *locally equicontinuous semigroup* if $T(0) = I$, $T(t)T(s) = T(s+t)$, for all $s,t \geq 0$ and for all $s < \infty, \{T(t)\mid 0 \leq t \leq s\}$ is equicontinuous.

The operator A *generates* $\{T(t)\}_{t\geq 0}$ if

$$Ax = \lim_{t\to 0} \frac{1}{t}\left(T(t)x - x\right),$$

with maximal domain, that is, $\mathcal{D}(A)$ equals the set of all x for which the limit exists.

Definition 4.6. Suppose all solutions of the abstract Cauchy problem (0.1) are unique; that is, there are no nontrivial solutions of (0.1), when $x = 0$. We will denote by Z the *solution space for A*, which we define to be the set of all x for which the abstract Cauchy problem (0.1) has a mild solution.

We topologize Z with the family of seminorms

$$\|x\|_{a,b,j} \equiv \sup_{t\in[a,b]} \|u(t,x)\|_j, \ a,b \in \mathbf{Q}^+, j \in \mathbf{N},$$

where $t \mapsto u(t,x)$ is the unique mild solution of (0.1).

It is convenient to note that Z is topologically equivalent to a subspace of $C([0,\infty), X)$, via the embedding

$$(\Lambda z)(t) \equiv u(t,z). \tag{4.7}$$

Theorem 4.8. The space Z is a Frechet space and $T(t)x \equiv u(t,x)$ is a locally equicontinuous semigroup on Z generated by $A|_Z$.

Remark 4.9. When X is a Banach space, then there will, in general, be no *Banach* solution space, that is, no Banach space that contains all solutions of the abstract Cauchy problem, on which A generates a strongly continuous semigroup, because a strongly continuous semigroup on a Banach space is automatically exponentially bounded (see Example 5.8). We shall see later (Theorem 5.5) that, for fixed $\omega \in \mathbf{R}$, there exists a Banach space on which A generates a strongly continuous semigroup, which

27

will contain all x for which the abstract Cauchy problem has a unique $O(e^{t(\omega-\epsilon)})$ solution, for any $\epsilon > 0$. To get *all* solutions of the abstract Cauchy problem, even when X is a Banach space, we must pass to a Frechet space. Since our solution space is still a Frechet space when X is, this seems to be the natural setting for our results in this chapter.

Here is a simple example of an operator, A, for which Z contains $\mathcal{D}(A)$, but the abstract Cauchy problem has no nontrivial exponentially bounded solutions.

Example 4.10. Let $\Omega \equiv \{x + iy \,|\, x \geq 0, e^{x^2} \leq y \leq e^{2x^2}\}$, and let $X \equiv H^\infty(\Omega) \cap BC(\Omega)$, the set of all functions holomorphic in the interior of Ω, continuous and bounded on Ω.

Let $Af(z) \equiv zf(z)$, on X, with maximal domain.

If u is a mild solution of the abstract Cauchy problem (0.1), then

$$(u(t))(z) = e^{tz}(u(0)(z)),$$

which is exponentially bounded only when the support of $u(0)$ is contained in a left half-plane. Because Ω is not contained in a left half-plane, and $u(0)$ is holomorphic in the interior of Ω, this can happen only when $u(0) \equiv 0$.

Note that A generates the A^{-1}-regularized semigroup

$$W(t)f(z) \equiv \frac{1}{z}e^{tz}f(z),$$

since $\|W(t)\|^2 = \sup_{x \geq 0} \frac{e^{2tx}}{(x^2 + e^{2x^2})}$. Thus by Theorem 4.15, $\mathcal{D}(A) \subsetneq Z$.

Note that, in the following, we do not assume that $\mathcal{D}(A)$ is dense or $\rho(A)$ is nonempty. It is surprising that both these facts follow automatically, when X is a Banach space, as consequences of generating a strongly continuous semigroup.

Corollary 4.11. *The following are equivalent.*

(a) The operator A generates a locally equicontinuous semigroup.

(b) There exists a unique mild solution of the abstract Cauchy problem (0.1), for all $x \in X$.

When $\rho(A)$ is nonempty, we get a similar equivalence for strong solutions of (0.1).

28

Corollary 4.12. *Suppose $\rho(A)$ is nonempty. Then the following are equivalent.*

(a) *The operator A generates a locally equicontinuous semigroup.*

(b) *The abstract Cauchy problem has a unique solution, for all $x \in \mathcal{D}(A)$.*

Theorem 4.13. *Suppose all solutions of the abstract Cauchy problem are unique. Then the following are equivalent.*

(a) *The abstract Cauchy problem has a mild solution, for all $x \in Im(C)$.*

(b) *$[Im(C)] \hookrightarrow Z$.*

(c) *There exists a mild C-existence family for A.*

The mild C-existence family for A is then given by

$$W(t) \equiv T(t)C,$$

where $T(t)$ is as in Theorem 4.8.

When A generates a regularized semigroup, we may use it to characterize Z. Also the uniqueness of the solutions follows automatically (see Theorem 3.5).

It is interesting that the choice of C in the following is irrelevant, that is, we still obtain the same space, Z.

Proposition 4.14. *Suppose A generates a C-regularized semigroup $\{W(t)\}_{t \geq 0}$. Then*

$$Z = \{x \mid W(t)x \in Im(C), \forall t \geq 0, \text{ and } t \mapsto C^{-1}W(t)x \text{ is continuouous }\},$$

and

$$\|x\|_{a,b,j} = \sup_{t \in [a,b]} \|C^{-1}W(t)x\|_j,$$

for all $j \in \mathbf{N}, a, b \in \mathbf{Q}$.

Theorem 4.15. *Suppose X is a Banach space, $\rho(A)$ is nonempty, C is injective and $CA \subseteq AC$. Then the following are equivalent.*

(a) *The abstract Cauchy problem (0.1) has a unique solution, for all $x \in C(\mathcal{D}(A))$.*

(b) *The abstract Cauchy problem has a unique mild solution, for all $x \in Im(C)$.*

(c) *All solutions of the abstract Cauchy problem are unique and*

$$[Im(C)] \hookrightarrow Z.$$

(d) *The operator A generates a C-regularized semigroup.*

(e) *All solutions of the abstract Cauchy problem are unique and there exists a mild C-existence family for A.*

(f) *There exists a mild C-existence family for A, $\{W(t)\}_{t \geq 0}$, such that $W(t)A \subseteq AW(t)$, for all $t \geq 0$.*

The C-regularized semigroup generated by A is then given by

$$W(t) \equiv T(t)C,$$

where $T(t)$ is as in Theorem 4.8.

Examples of operators that generate C-regularized semigroups are $i\Delta$, on $L^p(\mathbf{R}^n), 1 \leq p \leq \infty$ (with $C \equiv (1 + i\Delta)^{-k}$, where k depends on p and n); $-\Delta$, on $L^p(\Omega)$ (with $C \equiv e^{-(\Delta)^2}$) and $\begin{bmatrix} 0 & I \\ -\Delta & 0 \end{bmatrix}$, on $L^p(\Omega) \times L^p(\Omega)$ (with $C \equiv e^{-(\Delta)^2}$) (see Chapters XI, VIII and IX). Note that the corresponding abstract Cauchy problems are the Schrödinger equation, the backwards heat equation and the Cauchy problem for the Laplace equation. Because these are so ill-posed, Theorem 4.15 might be surprising. Note that in the three examples just given, the image of C is dense in X.

Example 4.16. Let A be as in Example 3.2,

$$(Af)(x) \equiv xf(x),$$

with maximal domain, on $C_0(\mathbf{R})$). Then Z equals the set of all continuous $f : \mathbf{R} \to \mathbf{C}$ that decay more rapidly on $[0, \infty)$ than any exponential, that is, for all $t \geq 0$, there exists a constant M_t such that $|f(x)| \leq M_t e^{-tx}$, for all $x \geq 0$. The space Z is topologized by the family of seminorms

$$\|f\|_{a,b} \equiv \sup_{x \geq 0} |f(x)e^{bx}| + \sup_{x \leq 0} |f(x)e^{ax}|,$$

for all $a, b \in \mathbf{Q}^+$ (compare with Example 5.6).

Note that, for any continuous g, not vanishing on any interval, such that $x \mapsto e^{tx}g(x)$ is bounded on $\mathbf{R}$, for all $t \geq 0$, A generates a C_g-regularized semigroup

$$W_g(t)f(x) \equiv e^{tx}g(x)f(x),$$

30

where $C_g \equiv W_g(0)$. Thus $C^{-1}W(t)f(x) = e^{tx}f(x)$, so that the construction of Proposition 4.14 yields our Z, regardless of which g is chosen.

It is clear how to extend this construction to arbitrary multiplication operators, A.

Example 4.17. Let $X \equiv C_0(\mathbf{R}) \bigcap C_0^1([0,\infty))$, where

$$C_0^1([0,\infty)) \equiv \{f \in C^1([0,\infty)) \mid \lim_{x \to \infty} f'(x) = 0 = \lim_{x \to \infty} f(x)\},$$

with the norm on X given by

$$\|f\| \equiv \|f\|_{C_0(\mathbf{R})} + \|f'\|_{C_0([0,\infty))}.$$

Let $A \equiv -\frac{d}{dx}$, with maximal domain, $\mathcal{D}(A) \equiv \{f \in X \mid f' \in X\}$.

We call X a *bumpy translation space*; it is good for counterexamples and other novel behaviour (see Examples 4.18, 5.8, 5.20, 19.11 and 28.6 and the Notes to Chapter IV). Behaviour that is acceptable in the negative real axis may not be acceptable in the positive real axis (e.g., nondifferentiable continuity), thus right-translation, the semigroup that A wants to generate, cannot map X into itself.

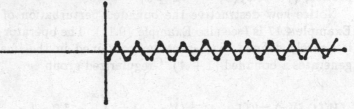

If the abstract Cauchy problem has a mild solution, it is given by right translation,
$$u(t,f)(x) = f(x - t), \quad (x \in \mathbf{R}, t \geq 0);$$

thus it is not hard to see that $Z = X \bigcap C^1(\mathbf{R})$, with topology generated by the seminorms

$$\|f\|_b \equiv \|f\|_{C_0(\mathbf{R})} + \sup_{x \geq -b} |f'(x)|,$$

for $b \in \mathbf{Q}^+$.

Example 4.18. Let X be the bumpy translation space of the previous example. Let A be a bounded perturbation of $-\frac{d}{dx}$,

$$(Af)(x) \equiv -f' + qf, \quad \mathcal{D}(A) \equiv \{f \in X \mid f' \in X\},$$

31

where $q \in X$, and q is not differentiable at any point in $(-\infty,0)$. Note that all right translates of q, $x \mapsto q(x-t)\,(t>0)$, fail to be in X.

A calculation shows that, if u is a mild solution of the abstract Cauchy problem, with $u(0,f)=f$, then

$$u(t,f)(x) = \exp\left[\int_0^t q(x-s)\,ds\right] f(x-t) \quad (x \in \mathbf{R}, t \geq 0).$$

Note that

$$\frac{d}{dx}[u(t,f)(x)] = (q(x) - q(x-t))\exp\left[\int_0^t q(x-s)\,ds\right] f(x-t)$$

$$+ \exp\left[\int_0^t q(x-s)\,ds\right] f'(x-t) \quad (x \in \mathbf{R}, t \geq 0);$$

for $x,t > 0, x-t < 0$, this will be defined if and only if $f(x-t)=0$. Thus, since $u(t,f)$ must be in $C_0^1([0,\infty))$, it follows that $f \equiv 0$ in $(-\infty,0)$.

We conclude that $Z = \{f \in C_0^1([0,\infty)) \,|\, f(0) = 0\}$, or, to be more precise, $\{f \in X \,|\, f(x) = 0 \;\forall x \leq 0\}$.

Notice how destructive the bounded perturbation of the operator in Example 4.17 is (see also Example 19.11). The operator of Example 4.17 is densely defined, has spectrum contained in the imaginary axis and generates a bounded $(1+A)^{-1}$-regularized group

$$(W(t)f)(x) \equiv ((1+A)^{-1}f)(x-t) \equiv \int_0^\infty e^{-s} f(x+s-t)\,ds \quad (x,t \in \mathbf{R})$$

and an exponentially bounded integrated semigroup (see Chapter XVIII)

$$(S(t)f)(x) \equiv \int_0^t f(x-s)\,ds \quad (x \in \mathbf{R}, t \geq 0).$$

Example 4.19. Suppose $(-\infty,0] \subseteq \rho(B)$, with $\{(1+r)\|(r+B)^{-1}\| \,|\, r \geq 0\}$ bounded. This includes, but is not limited to, all operators, B, such that $-B$ generates an exponentially decaying strongly continuous holomorphic semigroup (in fact, it may be shown to equal all squares of such operators). It is well known that one may construct *fractional powers*, $\{B^t\}_{t \in \mathbf{R}}$, for such operators, and that $\{B^{-t}\}_{t \geq 0}$ is a strongly continuous semigroup of bounded operators. Let $-A$ be the generator (we think of A as $\log B$).

Then it may be shown that Z is an extension of the locally convex space $C^\infty(B)$, with seminorms

$$\|x\|_k \equiv \|B^k x\|,$$

in the sense that $C^\infty(B) \subseteq Z$, and the topology of Z, restricted to $C^\infty(B)$, is equivalent.

In Chapter XV, we shall see that the family of unbounded operators $\{B^t\}_{t\geq 0}$ may be "regularized," that is, there exists an injective bounded operator C such that $\{B^t C\}_{t\geq 0}$ is a C-regularized semigroup.

Proof of Theorem 4.8: We will show that $\Lambda(Z)$ (see (4.7)) is a closed subspace of $C([0,\infty), X)$.

Suppose $< z_n > \subseteq Z$ and $\Lambda z_n \to \phi$, in $C([0,\infty), X)$. Then $T(t)z_n \to \phi(t)$, uniformly on compact subsets of $[0,\infty)$. For any n,

$$T(t)z_n = A\left(\int_0^t T(s)z_n\, ds\right) + z_n;$$

thus, letting $n \to \infty$, and using the fact that A is closed and the convergence is uniform, we find that $\int_0^t \phi(s)\, ds \in \mathcal{D}(A)$ and

$$\phi(t) = A\left(\int_0^t \phi(s)\, ds\right) + \phi(0), \forall t \geq 0,$$

thus $\phi(0) \in Z$ and $\phi(t) = T(t)\phi(0)$, so that $\phi = \Lambda(\phi(0))$, as desired.

Since Λ is a homeomorphism from Z onto $\Lambda(Z)$, this shows that Z is a Frechet space.

To show that $T(t)$ maps Z into itself and is a semigroup, we must show that, if $x \in Z$ and $s > 0$, then $u(s, x) \in Z$, and

$$u(t + s, x) = u(t, u(s, x)). \tag{4.20}$$

Let

$$v(t) \equiv \int_0^t u(r + s, x)\, dr = \int_s^{t+s} u(r, x)\, dr.$$

Then

$$v'(t) = u(t + s, x) = A\left(\int_0^{t+s} u(r, x)\, dr\right) + x$$

$$= A\left(v(t) + \int_0^s u(r, x)\, dr\right) + x = A(v(t)) + u(s, x) - x + x.$$

33

Thus $v(t) = \int_0^t u(r, u(s, x)) \, dr$, proving (4.20).

For $a, b \in \mathbf{Q}^+$, $j \in \mathbf{N}$, $s > 0$,

$$\|T(s)x\|_{a,b,j} = \sup\{\|u(t,x)\|_j \mid t \in [a + s, b + s]\} \leq \|x\|_{c,d,j},$$

when $c, d \in \mathbf{Q}^+$, $c \leq a + s$ and $d \geq b + s$. Thus $T(s) \in B(Z)$. The strong continuity of $\{T(s)\}_{s \geq 0}$ follows from the fact that $s \mapsto u(s, x)$, from $[a, b]$ into X, is continuous, hence uniformly continuous, for any $x \in Z, a, b \in \mathbf{Q}^+$. Since Z is a Frechet space, the strong continuity implies that $\{T(t)\}_{t \geq 0}$ is locally equicontinuous (see [Komu]).

Let $\tilde{A}$ be the generator of $\{T(s)\}_{s \geq 0}$. If $x \in \mathcal{D}(\tilde{A})$, then for $t > 0$, define

$$x_t \equiv \frac{1}{t} \int_0^t u(s, x) \, ds.$$

Then $x_t \in \mathcal{D}(A) \cap \mathcal{D}(\tilde{A})$ and $Ax_t = \tilde{A}x_t$, for all $t > 0$. Since $x \in \mathcal{D}(\tilde{A})$, the maps $s \mapsto u(s, x)$ and $s \mapsto \tilde{A}(u(s, x)) = \tilde{A}T(t)x$, from $[0, \infty)$ into Z, are both continuous. This implies that $\tilde{A}x_t \to \tilde{A}x$ and $x_t \to x$, in Z, as $t \to 0$.

Thus, since $Z \hookrightarrow X$, and A is closed, $x \in \mathcal{D}(A)$ and $Ax = \tilde{A}x \in Z$.

This is saying that $\tilde{A} \subseteq A|_Z$. To see that $A|_Z \subseteq \tilde{A}$, suppose $x \in \mathcal{D}(A|_Z)$. Then both x and Ax are in Z. We claim that

$$u(t, x) = \int_0^t u(s, Ax) \, ds + x, \ \forall t \geq 0. \tag{4.21}$$

To see this, let $w(t) \equiv \int_0^t (t - s) u(s, Ax) \, ds + tx$. We need to show that w is a solution of (2.2).

Since A is closed,

$$Aw(t) = \int_0^t A \left(\int_0^s u(r, Ax) \, dr \right) \, ds + tAx = \int_0^t (u(s, Ax) - Ax) \, ds + tAx$$

$$= \int_0^t u(s, Ax) \, ds = w'(t) - x,$$

as desired, proving (4.21).

By (4.21), $u(t, x)$ is a continuously differentiable function of t, from $[0, \infty)$ into X and $u'(t, x) = u(t, Ax)$. If $a, b \in \mathbf{Q}^+, j \in \mathbf{N}$ and $t > 0$, then

$$\left\| \frac{1}{t}(T(t)x - x) - Ax \right\|_{a,b,j} = \sup_{s \in [a,b]} \left\| \frac{1}{t}(u(s + t, x) - u(s, x)) - u'(s, x) \right\|_j.$$

34

This converges to 0 as $t \to 0$, since the map $t \mapsto u(t, x)$ is continuously differentiable. This means that $x \in \mathcal{D}(\tilde{A})$, thus $\tilde{A} = A|_Z$, as desired. ∎

Proof of Corollary 4.11: (a) → (b). This is well known and straightforward by letting $u(t, x) \equiv e^{tA}x$; see Chapter III or any of the references on locally equicontinuous semigroups.

(b) → (a). It is clear that $Z \hookrightarrow X$. Since both X and Z are Frechet spaces, and $X = Z$ (as sets), this implies that their topologies are equivalent. Thus, by Theorem 4.8, $T(t)x \equiv u(t, x)$ is a locally equicontinuous semigroup on X, generated by A. ∎

When $\rho(A)$ is nonempty, the following lemma shows the relationship between strong solutions and mild solutions.

Lemma 4.22. *Suppose $r \in \rho(A)$ and $x \in \mathcal{D}(A)$. Then the following are equivalent.*

(a) *The function $w(t)$ is a solution of (0.1).*

(b) *The function $u(t) \equiv (r - A)w(t)$ is a mild solution of (0.1), with x replaced by $(r - A)x$.*

Proof: (a) → (b). For any $t \geq 0$, since A is closed, we may integrate both sides of (0.1), to obtain, for any $t \geq 0$,

$$w(t) = A\left(\int_0^t w(s)\, ds\right) + x,$$

thus $\int_0^t w(r)\, dr \in \mathcal{D}(A^2)$, so that we may apply $(r - A)$ to both sides, as follows:

$$(r - A)w(t) = A\left(\int_0^t (r - A)w(s)\, ds\right) + (r - A)x. \tag{4.23}$$

(b) → (a). We apply $(r - A)^{-1}$ to both sides of (4.23). Since $w(s) \in \mathcal{D}(A)$ and $s \mapsto Aw(s)$ is continuous and A is closed, we may take A inside the integral to obtain

$$w(t) = \int_0^t Aw(s)\, ds + x,$$

as desired. ∎

Proof of Corollary 4.12: (a) → (b) is straightforward, as in Corollary 4.11.

(b) → (a). Suppose $r \in \rho(A)$. For any $x \in X$, by Lemma 4.22, $u(t, x) \equiv (r - A)u(t, (r - A)^{-1}x)$ defines a unique mild solution of (0.1), thus (a) follows from Corollary 4.11 ∎

35

Proof of Theorem 4.13: (a) → (c). Define $W : X \to C([0,\infty), X)$ by

$$(Wx)(t) \equiv W(t)x \equiv T(t)Cx = (\Lambda Cx)(t),$$

where Λ is defined by (4.7).

To see that W is closed, suppose $x_n \to x$ and $Wx_n \to \phi$. Since $\Lambda(Z)$ is closed in $C([0,\infty), X)$ (this was shown in the proof of Theorem 4.8), $\phi(t) = T(t)z$, for some $z \in Z$. Thus

$$z = \phi(0) = \lim_{n\to\infty} (Wx_n)(0) = \lim_{n\to\infty} Cx_n = Cx,$$

since $C \in B(X)$. Thus $\phi = Wx$, as desired. Thus by the closed graph theorem, W is bounded, which clearly implies that $W(t) \in B(X)$, for all $t \geq 0$.

The other properties of a C-existence family for A all follow from the properties of $T(t)$.

(c) → (b). Fix $a, b \in \mathbf{Q}^+, j \in \mathbf{N}$. By the uniform boundedness principle, $\{W(t)\}_{t\in[a,b]}$ is equicontinuous, thus there exists $M_{a,b,j}$ and $j_1, j_2, \ldots j_k$ such that

$$\sup_{t\in[a,b]} \|W(t)x\|_j \leq M_{a,b,j} \sum_{i=1}^{k} \|x\|_{j_i}, \forall x \in X.$$

Suppose $x \in Im(C)$ and $Cy = x$. Then $\|x\|_{a,b,j} = \|Cy\|_{a,b,j} = \sup_{t\in[a,b]} \|W(t)y\|_j$, thus, taking infima over all y such that $Cy = x$, we conclude that

$$\|x\|_{a,b,j} \leq M_{a,b,j} \sum_{i=1}^{k} \|x\|_{[Im(C)],j_i}.$$

∎

Proof of Proposition 4.14: This follows from Theorem 3.13. ∎

Proof of Theorem 4.15: The equivalence of (f) and (d) follows from Theorems 3.4(d), 3.5(1) and 3.7(a) and Proposition 3.9.

By Theorem 4.13, (b), (c) and (e) are equivalent.

The equivalence of (a) and (b) follows from Lemma 4.22, since, if $r \in \rho(A)$, then by Lemma 4.22, $u(t, x) \equiv (r - A)u(t, (r - A)^{-1}x)$ is a mild solution of (0.1) if and only if $u(t, (r - A)^{-1}x)$ is a solution of (0.1), with x replaced by $(r - A)^{-1}x$, and, since $CA \subseteq AC$, $C(\mathcal{D}(A))$ equals $Im((r - A)^{-1}C)$.

The fact that (a), (b) and (e) imply (f) follows by noting that, by uniqueness and Lemma 4.22,

$$W(t)x = u(t, Cx) = (r - A)u(t, C(r - A)^{-1}x) = (r - A)W(t)(r - A)^{-1}x.$$

36

To see that (f) implies (e), suppose u is a mild solution of (0.1), with $x = 0$. Then $v(t) \equiv \int_0^t u(s)\,ds$ is a strong solution of (0.1). For any $s, t \geq 0, 0 = \frac{d}{ds}W(t-s)v(s)$, so that $Cv(t) = W(t)v(0) = 0$; since C is injective, this implies that $v \equiv 0$, so that $u \equiv 0$, as desired. ∎

V. EXPONENTIALLY BOUNDED (BANACH) SOLUTION SPACES

We have seen that, in general, given an operator A, on a Banach space X, there is no Banach space embedded in X, that contains all solutions of the abstract Cauchy problem (0.1), on which A generates a strongly continuous semigroup (see Remarks 4.9, 5.20 and Example 4.10). However, there is such a Banach space that contains all *bounded* solutions of the abstract Cauchy problem; more generally, for any $\omega \in \mathbf{R}$, there exists a Banach space that contains all $O(e^{\omega t})$ solutions of the abstract Cauchy problem, on which A generates a strongly continuous semigroup.

We prove "pointwise Hille-Yosida"-type theorems (Theorems 5.10, 5.13, 5.14 and 5.15), that is, conditions on the resolvent evaluated at a fixed point x, that guarantee that the abstract Cauchy problem has an exponentially bounded solution for that particular choice of initial data. These conditions can be easily translated into global Hille-Yosida-type theorems, including the well-known Hille-Yosida theorem (see Chapter XVII).

Of particular interest is a pointwise condition that involves only $(z - A)^{-1}x$, for z in a right half-plane (Theorem 5.15), rather than all powers of the resolvent, since this seems like a particularly simple sufficient condition to verify.

This chapter, along with the previous chapter, also shows the relationship between general C-existence families and exponentially bounded C-existence families. It is the same as the relationship between Frechet spaces and Banach spaces; that is, a C-existence family for A corresponds to a restriction of A to a Frechet space generating a strongly continuous semigroup, while an exponentially bounded C-existence family corresponds to a restriction to a Banach space generating a strongly continuous semigroup. More specifically, an existence family corresponds to the topology of uniform convergence on compact subsets of $[0,\infty)$, while a bounded uniformly continuous existence family corresponds to the topology of uniform convergence on $[0,\infty)$ (see the definition of Z, in Chapter IV, the definition of Z_ω, in this chapter, and Propositions 4.14 and 5.18).

Throughout this chapter, C will be in $B(X)$. We will write $[Im(C)]$ for the Banach space $Im(C)$ with norm

$$\|x\|_{[Im(C)]} \equiv \inf\{\|y\| \mid Cy = x\}.$$

Also, A will be a fixed closed operator, on a Banach space X.

Definition 5.1. We will denote by Z_ω the $O(e^{\omega t})$ *solution space*, the set of all x for which the abstract Cauchy problem (0.1) has a mild solution, u, such that $t \mapsto e^{-\omega t}u(t)$ is bounded and uniformly continuous.

Note that Z_ω contains all x for which the abstract Cauchy problem has an $O(e^{(\omega-\epsilon)t})$ mild solution, for some $\epsilon > 0$.

When A has no eigenvalues in (a, ∞), for some real a, then by Proposition 2.9, exponentially bounded mild solutions of the abstract Cauchy problem are unique. When x is in Z_ω, we will then denote by $u(t, x)$ the unique mild solution such that $t \mapsto e^{-\omega t} u(t, x)$ is bounded and uniformly continuous.

We then define the following norm on Z_ω:

$$\|x\|_{Z_\omega} \equiv \sup_{t \geq 0} e^{-\omega t} \|u(t, x)\|.$$

We will call Z_0 the *Hille-Yosida space* for A, as in [K].

Since Z_ω is merely the Hille-Yosida space for $(A - \omega)$, it is sufficient to state our results for $\omega = 0$; we will leave it to the reader to make the translation to arbitrary ω.

Definition 5.2. We will denote by Y the *weak bounded solution space* or the *weak Hille-Yosida space* for A, the set of all x for which

$$\frac{d}{dt} w(t, x) = A\left(w(t, x)\right) + tx \ (t \geq 0), \ w(0, x) = 0, \qquad (5.3)$$

has a solution such that w' is Lipschitz continuous.

Note that $Z_0 \subseteq Y$. If $x \in Z_0$, then $w(t, x)$ is obtained from $u(t, x)$ by integrating twice, $w(t, x) = \int_0^t (t - s) u(s, x) \, ds$.

Here is the justification for our terminology (see also Theorem 5.5(3), (4) and (6)). Let $v(t, x) \equiv w'(t, x)$. When $\mathcal{D}(A)$ is dense, so that A^* is defined, then for any $x^* \in \mathcal{D}(A^*)$, the map $t \mapsto < x^*, v(t, x) >$ is Lipschitz continuous, thus is differentiable almost everywhere, with

$$\frac{d}{dt} < x^*, v(t, x) > = < A^* x^*, v(t, x) > + < x^*, x > \ a.e. \ (t \geq 0),$$

thus we are getting a "weak mild" solution of the abstract Cauchy problem.

As with Proposition 2.9, when there exists $\omega \in \mathbf{R}$ such that A has no eigenvalues in (ω, ∞), then the Lipschitz continuous solution of (5.3) is unique, so that we may define a norm on Y:

$$\|x\|_Y \equiv \sup_{r > s \geq 0} \{ \frac{1}{r - s} \|w'(r, x) - w'(s, x)\|, \|x\| \};$$

this may also be expressed as

$$\|x\|_Y = ess \sup_{\|x^*\| \leq 1, t \geq 0} \{ |\frac{d}{dt} < x^*, w'(t, x) > |, |x^*(x)| \}.$$

39

Note that $\|x\|_Y = \|x\|_{Z_0}$, for all $x \in Z_0$, since $\frac{d}{dt} < x^*, w'(t,x) >$ $=< x^*, u(t,x) >$ almost everywhere, for all $x \in Z_0, x^* \in X^*$.

Definition 5.4. We will say that an operator, B, satisfies the *Hille-Yosida conditions* if $(0,\infty) \subseteq \rho(B)$ and $\|s(s-B)^{-1}\| \leq 1$, for all $s > 0$.

Theorem 5.5. *Suppose A has no eigenvalues in $(0,\infty)$. Then*

(1) $Y \hookrightarrow X$;

(2) Z_0 *and Y are Banach spaces;*

(3) Z_0 *is the closure, in Y, of $\mathcal{D}(A|_Y)$;*

(4) $A|_Y$ *satisfies the Hille-Yosida conditions;*

(5) $T(t)x \equiv u(t,x)$ *is a strongly continuous semigroup on Z_0, generated by $A|_{Z_0}$, such that $\|T(t)\| \leq 1$, for all $t \geq 0$;*

(6) $\mathcal{D}(A|_Y)$ *equals the set of all x in $\mathcal{D}(A)$ for which the abstract Cauchy problem (0.1) has a bounded Lipschitz continuous mild solution;*

(7) $\mathcal{D}(A|_{Z_0})$ *equals the set of all x for which the abstract Cauchy problem has a bounded (strong) solution with a bounded uniformly continuous derivative;*

(8) Z_0 *is maximal-unique, that is, if W satisfies (1), (2) and (5), then $W \hookrightarrow Z_0$.*

Example 5.6. Let A be as in Example 3.2,

$$(Af)(x) \equiv xf(x),$$

with maximal domain, on $C_0(\mathbf{R})$). Then Z_ω equals $C_0((-\infty, \omega])$, that is, the set of all $f \in C_0(\mathbf{R})$ supported on $(-\infty, \omega]$, with the same norm

$$\|f\|_{Z_\omega} = \|f\|_{C_0((-\infty,\omega])},$$

which is the original $C_0(\mathbf{R})$ norm restricted to functions supported on $(-\infty, \omega]$.

The assertion about the norm can be seen by a short calculation:

$$\|f\|_{Z_\omega} \equiv \sup_{t \geq 0, x \in \mathbf{R}} |f(x)e^{t(x-\omega)}|$$

$$= \sup_{t \geq 0, x \leq 0} |f(x+\omega)e^{tx}| = \sup_{x \leq 0} |f(x+\omega)|.$$

Thus here Z_ω is merely a closed subspace of the original space.

See Example 4.16, for the solution space, Z, of this operator. Even for this simple example,

$$Z \neq \bigcup_{\omega \in \mathbf{R}} Z_\omega;$$

the function $f(s) \equiv e^{-s^2} \in Z$, but is not in Z_ω, for any real ω. For a much more extreme example of the gap between Z and Z_ω, see Example 5.9; see also Example 5.20.

Remark 5.7. The growth condition of (5) is necessary for maximality, in Theorem 5.5. That is, in general, given a closed operator A, there exists no maximal Banach space such that the restriction of A to that space generates a strongly continuous semigroup; given any Banach space Y, such that $A|_Y$ generates a strongly continuous semigroup, we can find a bigger Banach space on which A generates a strongly continuous semigroup (although there does exist a maximal *Frechet* space; see Chapter IV). Here is an example.

Example 5.8. This is an example of an operator, A, on a Banach space X, such that X has no maximal continuously embedded Banach subspace on which A generates a strongly continuous semigroup (compare with Theorem 5.5(8) and the implicit maximality of Theorem 4.8).

Let X be the bumpy translation space of Example 4.17, $X \equiv C_0(\mathbf{R}) \cap C_0^1([0, \infty))$, where

$$C_0^1([0, \infty)) \equiv \{f \in C^1([0, \infty)) \cap C_0([0, \infty)) \mid \lim_{x \to \infty} f'(x) = 0\}.$$

Let $A \equiv -\frac{d}{dx}$, with maximal domain, so that

$$e^{tA} f(x) = f(x - t),$$

right translation, for $f \in Z$.

For this operator A, $Z_0 = X \cap BC^1(\mathbf{R})$ (compare with Example 4.17), where $BC^1(\mathbf{R})$ is the set of bounded continuous functions with bounded continuous derivatives, on the real line.

Note first that, if W is any Banach space such that $A|_W$ generates a strongly continuous semigroup and $W \hookrightarrow X$, then there exists $\omega \in \mathbf{R}$ such that $W \hookrightarrow Z_\omega$. Mainly, choose any ω such that $e^{-\omega t} \|e^{tA|_W}\|$ is bounded (this is true for any A).

It is clear that $Z_{\omega_1} \hookrightarrow Z_{\omega_2}$, whenever $\omega_1 < \omega_2$. To show that we have the desired example, it is sufficient to show that, when $\omega_1 < \omega_2$, then $Z_{\omega_1} \neq Z_{\omega_2}$.

41

Define, for $\omega_1 < \omega < \omega_2$,

$$f(x) \equiv \sin(e^{-\omega x})1_{(-\infty,0]} + g(x)1_{(0,\infty)},$$

where g is chosen so that $f \in X$.

Then, for $t \geq 0, x \in \mathbf{R}$

$$\frac{d}{dx}(e^{tA}f)(x) = g'(x - t)1_{(t,\infty)} - \omega e^{-\omega(x-t)}\cos(e^{-\omega(x-t)})1_{(-\infty,t]};$$

in particular,

$$|\frac{d}{dx}(e^{tA}f)(0)| = \omega e^{\omega t}\cos(e^{\omega t}).$$

Thus $f \in Z_{\omega_2}$, but is not in Z_{ω_1}.

Example 5.9. Let A be as in Example 4.10. Then, for all $\omega \in \mathbf{R}$, Z_ω is trivial, but Z contains $\mathcal{D}(A)$.

The equivalence of (a) and (b) in the following may be considered a "pointwise Hille-Yosida theorem." It is not hard to see how the well-known Hille-Yosida theorem, for strongly continuous semigroups, follows from this. More generally, in Theorem 17.4 (see also Theorem 17.3), we use this to obtain a Hille-Yosida-type characterization of densely defined generators of exponentially bounded C-regularized semigroups; choosing $C = I$ then gives the usual Hille-Yosida theorem.

Theorem 5.10. *Suppose A has no eigenvalues in $(0,\infty)$. Then the following are equivalent:*

(a) $x \in Y$; and

(b) $x \in Im(s - A)^n$, for all $s > 0, n \in \mathbf{N}$ and

$$\{s^n\|(s - A)^{-n}x\| \mid s > 0, n \in \mathbf{N}\}$$

is bounded.

Then

$$\|x\|_Y = \sup\{s^n\|(s - A)^{-n}x\| \mid s > 0, n + 1 \in \mathbf{N}\}.$$

When A has no eigenvalues in a right half-plane, we may represent solutions of the abstract Cauchy problem as complex integrals of the (possibly unbounded) resolvent, and also obtain characterizations of Z_ω or Y.

Definition 5.11. Suppose $a \in \mathbf{R}$ and A has no eigenvalues in $Re(z) > a$. Let $\mathcal{D}_a$ be the set of all $x \in \bigcap_{Re(z)>a} Im(z - A)$ such that the map $z \mapsto (z - A)^{-1}x$ is holomorphic and bounded on $Re(z) > a$.

Then we define, for any $t \geq 0$, $x \in \mathcal{D}_a, \omega > a$,

$$K(t)x \equiv \int_{\omega + i\mathbf{R}} e^{zt}(z - A)^{-1}x \, \frac{dz}{2\pi i z^2}.$$

Note that, by the residue theorem, this is independent of $\omega > a$.

Even when A itself generates a bounded strongly continuous semigroup, we only expect $\|(z - A)^{-1}\|$ to be *bounded* in left half-planes, rather than $O(|z|^{-1})$, unless the semigroup is holomorphic.

For $y \in \mathcal{D}_a$, we obtain solutions of the three-times integrated abstract Cauchy problem.

Proposition 5.12. *Suppose $y \in \mathcal{D}_a$, for some $a \in \mathbf{R}$. Then, for all $t \geq 0$, $\int_0^t K(s)y \, ds \in \mathcal{D}(A)$ and*

$$A\left(\int_0^t K(s)y \, ds\right) = K(t)y - \frac{1}{2}t^2 y.$$

Theorem 5.13. *Suppose A has no eigenvalues in the open right half-plane $Re(z) > 0$. Then the following are equivalent:*

(a) *$x \in Y$; and*

(b) *$x \in \mathcal{D}_a$, for all $a > 0$, and the map $t \mapsto K(t)x$, from $[0, \infty)$ into X, has a Lipschitz continuous derivative.*

Then $K(t)x = w(t, x)$, from Definition 5.2.

Theorem 5.14. *Suppose A has no eigenvalues in the open right half-plane $Re(z) > 0$. Then the following are equivalent:*

(a) *$x \in Z_0$; and*

(b) *$x \in \mathcal{D}_a$, for all $a > 0$, and the map $t \mapsto K(t)x$, from $[0, \infty)$ into X, has a bounded uniformly continuous second derivative.*

Then $K''(t)x = u(t, x)$, from Definition 5.1.

The following is a simpler sufficient condition for being in the Hille-Yosida space, than appears in Theorem 5.10, because it involves only the resolvents, $(z - A)^{-1}$, rather than arbitrarily high powers of the resolvents; the only possible increase in difficulty is that we need the resolvents in a half-plane $Re(z) > a$, rather than a half-line (w, ∞). We also should point out that this is not a necessary condition, as was the case in Theorem 5.10.

Theorem 5.15. *Suppose $k \in \mathbf{N}, \delta > 0$, A has no eigenvalues in $Re(z) > \delta$, $x \in \mathcal{D}(A^{k+1})$, $A^{k+1}x \in \bigcap_{Re(z)>\delta} Im(z - A)$ and the map $z \mapsto (z - A)^{-1}A^{k+1}x$ is holomorphic and $O(|z|^{k-1})$ in $Re(z) > \delta$.*

Then $x \in Z_b$, for all $b > \delta$.

The following three results are proven exactly as are Theorems 4.13 and 4.15 and Proposition 4.14.

Theorem 5.16. *Suppose A has no eigenvalues in $(0, \infty)$. Then the following are equivalent.*

(a) *The abstract Cauchy problem (0.1) has a mild bounded uniformly continuous solution, for all $x \in Im(C)$.*

(b) *$[Im(C)] \hookrightarrow Z_0$.*

(c) *There exists a bounded, strongly uniformly continuous mild C-existence family for A.*

The mild C-existence family for A is then given by

$$W(t) = e^{tA|z_0}C.$$

Theorem 5.17. *Suppose $CA \subseteq AC$ and $\rho(A)$ is nonempty. Then the following are equivalent.*

(a) *The abstract Cauchy problem (0.1) has a unique solution, for all $x \in C(\mathcal{D}(A))$, with u and u' bounded and uniformly continuous functions from $[0, \infty)$ into X.*

(b) *The abstract Cauchy problem has a mild bounded uniformly continuous solution, for all $x \in Im(C)$.*

(c) *$[Im(C)] \hookrightarrow Z_0$.*

(d) *All solutions of the abstract Cauchy problem are unique and there exists a bounded, strongly uniformly continuous mild C-existence family for A.*

(d) *The operator A generates a bounded, strongly uniformly continuous C-regularized semigroup.*

Proposition 5.18. *Suppose A generates a bounded strongly uniformly continuous C-regularized semigroup $\{W(t)\}_{t \geq 0}$. Then*

$$Z_0 = \{x \mid t \mapsto C^{-1}W(t)x \text{ is uniformly continuous and bounded}\}$$

and

$$\|x\|_{Z_0} = \sup_{t \geq 0} \|C^{-1}W(t)x\|.$$

Remark 5.19. If A is densely defined and generates a bounded C-regularized semigroup $\{W(t)\}_{t\geq 0}$, then $\{W(t)\}_{t\geq 0}$ is strongly uniformly continuous. The same is true if $\{W(t)\}_{t\geq 0}$ is stable. It is not clear if all bounded regularized semigroups are strongly uniformly continuous.

Remark and Example 5.20. The following example shows that, even when A generates a bounded strongly uniformly continuous regularized semigroup, as in Theorem 5.17 (and an exponentially bounded once-integrated semigroup; see Chapter XVIII and Example 4.18), $\bigcup_{\omega \in \mathbf{R}} Z_\omega$ may not contain all initial data for which the abstract Cauchy problem has a solution, that is, we may have

$$Z \neq \bigcup_{\omega \in \mathbf{R}} Z_\omega.$$

Let X be the bumpy translation space $X \equiv C_0(\mathbf{R}) \cap C_0^1([0,\infty))$, as in Example 4.17, $A \equiv -\frac{d}{dx}, \mathcal{D}(A) \equiv \{f \in X | f' \in X\}$ (see also Example 5.8). Then, as shown in Example 4.18, A generates a bounded strongly uniformly continuous $(1 + A)^{-1}$-regularized semigroup.

If $x \in \bigcup_{\omega \in \mathbf{R}} Z_\omega$, then $u(t,x)$ would be exponentially bounded. However, there are solutions of the abstract Cauchy problem that are not exponentially bounded. Choose

$$f(x) \equiv \sin(e^{x^2})1_{(-\infty,0]} + g(x)1_{[0,\infty)},$$

where g is chosen so that $f \in X$ and $\|f\| = 1$. Then $u(t,f)(x) \equiv f(x - t)$ is the solution of the abstract Cauchy problem and $\|u(t,f)\| \geq e^{t^2}$, for all $t \geq 0$.

We have already seen that, to obtain a space on which A generates a strongly continuous semigroup, containing all initial data for which the abstract Cauchy problem has a mild solution, we must allow Frechet spaces. This example shows that this is true even when A generates a bounded strongly uniformly continuous C-regularized semigroup, with the image of C as large as the domain of A.

In this example, Z_0 equals

$$\{f \in C_0^1([0,\infty)) \cap C^1(\mathbf{R}) | f' \text{ is bounded and uniformly continuous on} \mathbf{R}\},$$

with $\|f\|_{Z_0} \equiv \sup_{x \in \mathbf{R}} |f(x)| + \sup_{x \in \mathbf{R}} |f'(x)|$.

More generally, if $X \equiv \bigcap_{k=0}^n C_0^k([k,\infty))$ with

$$\|f\| \equiv \sum_{k=0}^n \sup_{x \geq k} |f^{(k)}(x)|$$

45

and $A \equiv -\frac{d}{dx}, D(A) \equiv \{f \in X \mid f' \in X\}$, then A generates a bounded strongly uniformly continuous $(1 - A)^{-n}$-regularized semigroup, and Z_0 equals $C_0^n([0,\infty))$, $\|f\|_{Z_0} = \sum_{k=0}^{n} \sup_{x \geq 0} |f^{(k)}(x)|$.

Proof of Theorem 5.5: Assertion (1) is obvious from the definition of $\|x\|_Y$.

The fact that Z_0 is a Banach space and $\{T(t)\}_{t \geq 0}$ is a strongly continuous semigroup generated by $A|_{Z_0}$ follows as in Chapter IV, Theorem 4.8; note that the *uniform* continuity of the solutions implies the strong continuity of $\{T(t)\}_{t \geq 0}$. For any $t \geq 0$, $x \in Z_0$,

$$\|T(t)x\|_{Z_0} = \sup_{s \geq 0} \|u(t+s,x)\| = \sup_{r \geq t} \|u(r,x)\| \leq \|x\|_{Z_0}.$$

To see that Y is a Banach space, suppose $< x_n >$ is a Cauchy sequence in Y. Then there exists Lipschitz continuous $v : [0,\infty) \to X$ such that the maps $t \mapsto w'(t, x_n)$ converge to v in the Lipschitz norm, as $n \to \infty$. Let $w(t) \equiv \int_0^t v(s)\, ds$, for $t \geq 0$. There exists $x \in X$ such that $x_n \to x$, in X, as $n \to \infty$. Since A is closed and, for all $n \in \mathbf{N}, t \geq 0$,

$$w'(t, x_n) = A(w(t, x_n)) + tx_n,$$

it follows that $w(t) \in \mathcal{D}(A)$ and $w'(t) = A(w(t)) + tx$, for all $t \geq 0$. Thus $x \in Y$, with $w(t, x) = w(t)$, so that $x_n \to x$ in Y, as $n \to \infty$.

This proves (2) and (5).

For (4), fix $s > 0$. Note that, for $x \in Y$, the same proof as in Lemma 2.10 shows that

$$(s - A)\left(s \int_0^\infty e^{-st} w'(t,x)\, dt\right) = (s - A)\left(s^2 \int_0^\infty e^{-st} w(t,x)\, dt\right) = x,$$

$$(5.21)$$

for all $s > 0$; note that the quantities in parentheses are in $\mathcal{D}(A)$ because A is closed.

Thus $Y \subseteq Im(s - A) = \mathcal{D}((s - A)^{-1})$. To show that $(s - A)^{-1}$ maps Y into itself, for a fixed $x \in Y$, we must construct a map $r \mapsto w(r, (s - A)^{-1}x)$, satisfying (5.3). A candidate is suggested by the formal identity $e^{rA}(s - A)^{-1} = \int_0^\infty e^{-st} e^{(r+t)A}\, dt$, which, after integrating both sides with respect to r twice, and thinking of $w''(r)$ as being e^{rA}, gives us the following definition.

$$v(r) \equiv \int_0^\infty e^{-st} \left[w(r+t,x) - w(t,x) - rw'(t,x) \right] dt.$$

46

Then
$$v'(r) = \int_0^\infty e^{-st} \left[w'(r+t,x) - w'(t,x) \right] dt,$$

and

$$Av(r) = \int_0^\infty e^{-st} \left[w'(r+t,x) - (r+t)x - (w'(t,x) - tx) \right] dt$$

$$- rA \int_0^\infty e^{-st} w'(t,x)\, dt,$$

so that, by (5.21),

$$v'(r) - Av(r) = \frac{r}{s}(x + A(s-A)^{-1}x) = \frac{r}{s}(x + ((A-s)+s)(s-A)^{-1}x)$$

$$= r(s-A)^{-1}x.$$

This proves that $(s-A)^{-1}$ maps Y into itself, with $w(r,(s-A)^{-1}x) = v(r)$, for $x \in Y$.

This implies that, for all $s > 0$, $(s - A|_Y)$ is a bijection, with

$$\frac{d}{dr} < x^*, w'(r,(s-A|_Y)^{-1}x) >= \int_0^\infty e^{-st} \frac{d}{dr} < x^*, w'(r+t,x) > dt,$$

(5.22)

for all $r, s > 0, x \in Y, x^* \in X^*$.

To prove the desired estimates on $\|(s - A|_Y)^{-1}\|$, we will use the last expression for $\|x\|_Y$, in Definition 5.2.

By (5.22), for $s > 0$,

$$\text{ess} \sup_{t \geq 0, \|x^*\| \leq 1} |\frac{d}{dt} < x^*, w'(t,(s-A|_Y)^{-1}x) >|$$

$$\leq \frac{1}{s} \text{ess} \sup_{t \geq 0, \|x^*\| \leq 1} |\frac{d}{dt} < x^*, w'(t,x) >| \leq \frac{1}{s}\|x\|_Y.$$

Also, by (5.21),

$$| < x^*, (s-A|_Y)^{-1}x > | = |\int_0^\infty e^{-st} \frac{d}{dt} < x^*, w'(t,x) > dt| \leq \frac{1}{s}\|x\|_Y,$$

by the same argument.

Thus

$$\|s(s-A|_Y)^{-1}x\|_Y \leq \|x\|_Y,$$

47

for $x \in Y, s > 0$, as desired, proving (4).

To prove (3), suppose $x \in \mathcal{D}(A|_Y)$. By (4), there exists $y \in Y$ such that $x = (1 - A)^{-1}y$. Thus $w'(t, x) = (1 - A)^{-1}w'(t, y) \in \mathcal{D}(A)$; letting $v \equiv w'$, we may now rewrite (5.3) as

$$v(t, x) = A\left(\int_0^t v(s)\,ds\right) + tx = \int_0^t Av(s)\,ds + tx,$$

so that v is a mild solution of (0.1), that is, $x \in Z_0$. Since Z_0 is a closed subspace of Y, this implies that the closure of $\mathcal{D}(A|_Y)$ is contained in Z_0. Conversely, since the generator of a strongly continuous semigroup is densely defined (see Theorem 3.4(b)), and we've already shown (5), it follows that Z_0 equals the closure of $\mathcal{D}(A|_{Z_0})$, which is contained in the closure of $\mathcal{D}(A|_Y)$. This proves (3).

For (6), suppose $x \in \mathcal{D}(A|_Y)$. Then both x and Ax are in Y and, by (3), $x \in Z_0$. Thus $u(t, x)$ is a bounded mild solution of (0.1). We will show that $w'(t, Ax) = u(t, x) - x$, which, by the definition of Y, will imply that $u(t, x)$ is Lipschitz continuous.

Let $w(t) \equiv \int_0^t u(s, x)\,ds - tx$. Then by the definition of a mild solution, $w(t) \in \mathcal{D}(A)$, for all $t \geq 0$, with $Aw(t) = u(t, x) - x - tAx$, so that $w'(t) - Aw(t) = tAx$. This implies that $w(t) = w(t, Ax)$, as desired.

Conversely, suppose (0.1) has a bounded Lipschitz continuous mild solution, $u(t, x)$, and $x \in \mathcal{D}(A)$. Since we already know that $x \in Z_0 \subseteq Y$, all that we need show is that $Ax \in Y$; this follows by showing that $w(t, Ax) \equiv \int_0^t u(s, x)\,ds - tx$ is the desired solution of (5.3).

For (7), suppose $x \in \mathcal{D}(A|_{Z_0})$. Since x and Ax are in Z_0, both $u(t, x)$ and $u(t, Ax)$ are bounded and uniformly continuous. As in the proof of (6), $u(t, x) = \int_0^t u(s, Ax)\,ds + x$, so that $t \mapsto u(t, x)$ is a bounded strong solution of (0.1), with a bounded uniformly continuous derivative $u(t, Ax)$.

Conversely, suppose (0.1) has a bounded (strong) solution with a bounded uniformly continuous derivative. Then $x \in Z_0 \cap \mathcal{D}(A)$, so all that remains is to show that $Ax \in Z_0$. Let $u(t) \equiv u'(t, x)$. Then it is simple to verify that u is a mild solution of (0.1), with initial data Ax, that is, $A(\int_0^t u(s)\,ds) = u(t) - Ax$, as desired.

For (8), suppose W satisfies (1), (2) and (5). Since $W \hookrightarrow X, u(t, x) \equiv e^{tA|_W}x$ is the unique bounded mild solution of (0.1) and is uniformly continuous. Thus $W \subseteq Z_0$. There exists K such that

$$\|x\| \leq K\|x\|_W, \forall x \in W.$$

For any $t \geq 0$,

$$\|u(t, x)\| = \|e^{tA|_W}x\| \leq K\|e^{tA|_W}x\|_W \leq K\|x\|_W,$$

thus $\|x\|_{Z_0} \leq K\|x\|_W$, for all $x \in W$, which is saying that $W \hookrightarrow Z_0$. ∎

Lemma 5.23. *Suppose A has no eigenvalues in $(0,\infty)$, $x \in Im(s - A)^n$, for all $s > 0, n \in \mathbf{N}$ and there exists $M < \infty$ such that*

$$\|s^n(s - A)^{-n}x\| \le M\|x\|, \forall n \in \mathbf{N}, s > 0.$$

Then the map $s \mapsto (s - A)^{-1}x$, from $(0,\infty)$ into X, is infinitely differentiable, with

$$(\frac{d}{ds})^n((s - A)^{-1}x) = (-1)^n n!(s - A)^{-(n+1)}x.$$

Proof: Fix $r > 0$. For complex s such that $|r - s| < r$, define

$$R(s) \equiv \sum_{k=0}^{\infty}(r - s)^k(r - A)^{-(k+1)}x. \tag{5.24}$$

We will show that $R(s) = (s - A)^{-1}x$, when s is real. The hypotheses on A imply that the series in (5.24) converges uniformly, in the norm of X, on $\{s | |s| \le s_0\}$, for any $s_0 < r$. Thus $R(s)$ is in X. Note that each partial sum in (5.24) is in $\mathcal{D}(A)$, with

$$A\left(\sum_{k=0}^{n}(r - s)^k(r - A)^{-(k+1)}x\right) = \sum_{k=0}^{n}\left[A(r - A)^{-1}\right](r - s)^k(r - A)^{-k}x.$$

The hypotheses on A guarantee that the latter sum converges, as $n \to \infty$, thus, since A is closed, $R(s) \in \mathcal{D}(A)$, with

$$AR(s) = \sum_{k=0}^{\infty}(r - s)^k A(r - A)^{-(k+1)}x.$$

Thus we may calculate as follows.

$$(s - A)R(s) = \sum_{k=0}^{\infty}(r - s)^k\left[(r - A) - (r - s)\right](r - A)^{-(k+1)}x$$

$$= \sum_{k=0}^{\infty}(r - s)^k\left[I - (r - s)(r - A)^{-1}\right](r - A)^{-k}x$$

$$= x.$$

Thus $x \in Im(s - A)$ and $(s - A)^{-1}x$ equals $R(s)$, when s is real and $|r - s| < r$. This implies that $r \mapsto (r - A)^{-1}x$, from $(0,\infty)$ into X, is analytic, with

$$\left(\frac{d}{dr}\right)^n\left[(r - A)^{-1}x\right] = \left(\frac{d}{ds}\right)^n R(s)|_{s=r}$$

$$= (-1)^n n!(r - A)^{-(n+1)}x. \quad \blacksquare$$

Lemma 5.25 (from [A2]). *Suppose W is a Banach space, $M < \infty, g :$ $(0, \infty) \to W$. Then the following are equivalent.*

(a) *The function g is the once-integrated Laplace transform of a Lipschitz continuous function of Lipschitz constant M, that is, there exists $G : [0, \infty) \to W$ such that $G(0) = 0, \|G(t) - G(s)\| \leq M|t - s|$, for all $s, t \geq 0$ and*

$$g(s) = s \int_0^\infty e^{-st} G(t)\, dt, \ (s > 0).$$

(b) *The function g is infinitely differentiable and*

$$\left\| \frac{s^{n+1}}{n!} g^{(n)}(s) \right\| \leq M, \ \forall s > 0, n + 1 \in \mathbf{N}.$$

Proof of Theorem 5.10: (a) $\to$ (b). By Theorem 5.5(4),

$$s^n \|(s - A)^{-n}\|_Y \leq 1,$$

for all $s > 0, n \in \mathbf{N}$, thus (b) follows from the fact that $Y \hookrightarrow X$.

(b) $\to$ (a). By Lemmas 5.23 and 5.25, there exists Lipschitz continuous $G(t, x)$ such that

$$(s - A)^{-1}x = s \int_0^\infty e^{-st} G(t, x)\, dt, \tag{5.26}$$

for all $s > 0$.

Since $A(s - A)^{-1}x = s(s - A)^{-1}x - x$, we may apply these same lemmas to $g(s) \equiv \frac{1}{s}(s - A)^{-1}x$, on the Banach space (since A is closed) $[\mathcal{D}(A)]$, to obtain Lipschitz continuous $t \mapsto G_A(t, x)$, from $[0, \infty)$ to $[\mathcal{D}(A)]$, such that

$$\frac{1}{s}(s - A)^{-1}x = s \int_0^\infty e^{-st} G_A(t, x)\, dt,$$

for all $s > 0$. However, applying integration by parts to (5.26) gives us

$$\frac{1}{s}(s - A)^{-1}x = s \int_0^\infty e^{-st} \int_0^t G(r, x)\, dr\, dt,$$

for all $s > 0$. By the uniqueness of the Laplace transform, $G_A(t, x) = \int_0^t G(r, x)\, dr$, for all $t \geq 0$.

50

This implies that $\int_0^t G(r,x)\,dr \in \mathcal{D}(A)$, for all $t \geq 0$ and the map $t \mapsto A \int_0^t G(r,x)\,dr$, from $[0,\infty)$ into X, is continuous.

We will show that $w(t) \equiv \int_0^t G(r,x)\,dr$ is the desired solution of (5.3), by showing that the Laplace transforms of Aw and $w'(t) - tx$ are equal. For $s > 0$, since A is closed,

$$s \int_0^\infty e^{-st} Aw(t)\,dt = A \int_0^\infty e^{-st} G(t,x)\,dt$$

$$= (A - s + s)\frac{1}{s}(s - A)^{-1}x$$

$$= (s - A)^{-1}x - \frac{1}{s}x$$

$$= s \int_0^\infty e^{-st}(w'(t) - tx)\,dt.$$

This implies that $w(t) = w(t,x)$, the desired solution of (5.3), so that $x \in Y$.

For the assertion about $\|x\|_Y$, first note that we have shown that

$$w'(t,x) = G(t,x) \quad (t \geq 0, x \in Y),$$

so that

$$\|x\|_Y = \sup_{r > s \geq 0} \{\frac{1}{r - s}\|G(r,x) - G(s,x)\|, \|x\|\};$$

by Lemmas 5.25 and 5.23 (in that order),

$$\sup_{r > s \geq 0} \{\frac{1}{r - s}\|G(r,x) - G(s,x)\|\}$$

$$= \sup\{\|\frac{s^{n+1}}{n!}(\frac{d}{ds})^n \left((s - A)^{-1}x\right)\| \mid s > 0, n + 1 \in \mathbf{N}\}$$

$$= \sup\{s^n \|(s - A)^{-n}x\| \mid s > 0, n \in \mathbf{N}\},$$

concluding the proof. ∎

Proof of Proposition 5.12: By Fubini's theorem,

$$\int_0^t K(s)y\,ds = \int_{\omega + i\mathbf{R}} (e^{zt} - 1)(z - A)^{-1}y \frac{dz}{2\pi i z^3};$$

51

thus, since A is closed, the growth conditions on $\|(z-A)^{-1}y\|$ imply that this is in $\mathcal{D}(A)$, with

$$A\left(\int_0^t K(s)y\,ds\right) = \int_{\omega+i\mathbf{R}} (e^{zt}-1)(A-z+z)(z-A)^{-1}y\,\frac{dz}{2\pi i z^3}$$

$$= \left(\int_{\omega+i\mathbf{R}} (1-e^{zt})\,\frac{dz}{2\pi i z^3}\right)y$$

$$+ \int_{\omega+i\mathbf{R}} (e^{zt})(z-A)^{-1}y\,\frac{dz}{2\pi i z^2}$$

$$= K(t)y - \frac{1}{2}t^2 y,$$

by a calculus of residues argument. $\blacksquare$

Lemma 5.27. *Suppose* $x \in \mathcal{D}_a$, *for all* $a > 0$. *Then for all* $s > 0$,

$$\int_0^\infty e^{-st}K(t)x\,dt = \frac{1}{s^2}(s-A)^{-1}x.$$

Proof: For fixed $s > 0$, choose ω between s and 0. Then, by Fubini's theorem,

$$\int_0^\infty e^{-st}K(t)x\,dt = \int_{\omega+i\mathbf{R}} \int_0^\infty e^{t(z-s)}(z-A)^{-1}x\,dt\,\frac{dz}{2\pi i z^2}$$

$$= \int_{\omega+i\mathbf{R}} \frac{1}{s-z}(z-A)^{-1}x\,\frac{dz}{2\pi z^2},$$

which a calculus of residues argument shows to be equal to $\frac{1}{s^2}(s-A)^{-1}x$. $\blacksquare$

Proof of Theorem 5.13: (b) $\rightarrow$ (a). By Proposition 5.12, $K(t)x$ is a solution of (5.3), if we can show that $K(t)x \in \mathcal{D}(A)$.

This follows by using the fact that A is closed:

$K(t)x = \lim_{h\to 0} \frac{1}{h}\int_t^{t+h} K(s)x\,ds$ and $\lim_{h\to 0} A\left(\frac{1}{h}\int_t^{t+h} K(s)x\,ds\right) = K'(t)x - tx$, so $K(t)x \in \mathcal{D}(A)$ and

$$AK(t)x = K'(t)x - tx;$$

this shows that $x \in Y$, with $K(t)\ddot{x} = w(t,x)$.

(a) $\rightarrow$ (b). As in the proof of Lemma 2.10, $x \in Im(z-A)$ and

$$(z-A)^{-1}x = z\int_0^\infty e^{-zr}w'(r,x)\,dr,$$

52

when $Re(z) > 0$. Since w' is Lipschitz, this implies that $x \in \mathcal{D}_a$, for all $a > 0$.

Also, an integration by parts and Lemma 5.27 show that $K(t)x$ and $w(t, x)$ have the same Laplace transform, thus are equal, so that $K(t)x = w(t, x)$. ∎

Proof of Theorem 5.14: (b) → (a). As in the proof of Theorem 5.13 (b) → (a), $K(t)x \in \mathcal{D}(A)$, with

$$AK(t)x = K'(t)x - tx,$$

for all $t \geq 0$. Repeating the argument given there implies that $K'(t)x \in \mathcal{D}(A)$, with

$$AK'(t)x = K''(t)x - x,$$

so that $x \in Z_0$, with $u(t, x) = K''(t)x$.

(a) → (b). By Theorem 5.13, $x \in \mathcal{D}_0$ and

$$K(t)x = \int_0^t w'(r, x)\, dr = \int_0^t (t - r)u(r, x)\, dr,$$

so that $K(t)x$ is twice differentiable and $K''(t)x = u(t, x)$. ∎

Proof of Theorem 5.15: First note that, by writing

$$(z - A)^{-1} = \frac{1}{z}(1 + A(z - A)^{-1}),$$

it can be shown that $(z - A)^{-1}A^m x$ is $O(|z|^{k-1})$ in $Re(z) > \delta$, whenever $0 \leq m \leq k + 1$. Thus $(z - A)^{-1}(A - a)^{k+1}x$ is $O(|z|^{k-1})$, for any $a > \delta$.

For $a > \delta$, the resolvent identity implies that

$$(z - A)^{-1}(A - a)^{-n} = \sum_{i=0}^{n-1} \frac{1}{(z - a)^{i+1}}(A - a)^{i-n} + \frac{1}{(z - a)^n}(z - A)^{-1},$$

for all $n \in \mathbf{N}, Re(z) > a$; thus, for $j = 0, 1, 2$,

$$(z - A)^{-1}(A - a)^j x = \sum_{i=0}^{k-j} \frac{1}{(z - a)^{i+1}}(A - a)^{i+j}x$$

$$+ \frac{1}{(z - a)^{k+1-j}}(z - A)^{-1}(A - a)^{k+1}x.$$

53

Thus, for $a > \delta$, x, Ax and $A^2 x$ are in $\mathcal{D}_a$ (see Definition 5.11), so that we may apply Proposition 5.12, with $y = x, Ax, A^2 x$, as follows.

$$K(t)x - \frac{1}{2}t^2 x = A\left(\int_0^t K(s)x\, ds\right) = \int_0^t K(s)Ax\, ds, \qquad (5.28)$$

so that $K(t)x$ is differentiable, with

$$K'(t)x - tx = K(t)Ax = \frac{1}{2}t^2 Ax + A\left(\int_0^t K(s)Ax\, ds\right)$$

$$= \frac{1}{2}t^2 Ax + \int_0^t K(s)A^2 x\, ds, \qquad (5.29)$$

so that both $K'(t)x$ and $K(t)Ax$ are differentiable.

Now suppose $b > \delta$. Let $v(t) \equiv K''(t)x$. Differentiating both sides of (5.28), and using the fact that A is closed, now implies that v is a mild solution of (0.1); this implies that $u(t) \equiv e^{-bt}v(t)$ is a mild solution of

$$u'(t) = (A - b)u(t) + x, \; u(0) = 0.$$

All that remains is to show that u is bounded and uniformly continuous. By (5.29),

$$e^{bt}u(t) = x + tAx + K(t)A^2 x \; (t \geq 0).$$

By choosing ω such that $b > \omega > a > \delta$, we see that, since $\|K(t)A^2 x\|$ is $O(e^{\omega t})$ (see Definition 5.11), $u \in C_0([0, \infty), X)$, thus is bounded and uniformly continuous, as desired. ∎

VI. WELL-POSEDNESS ON A LARGER SPACE; GENERALIZED SOLUTIONS

Another way to approach ill-posed problems is to expand one's definition of solution, so that all initial data in the original space yield a solution, if one allows "generalized" solutions. A well-known example of this is the theory of distributions, generalizing the idea of a function.

This is in contrast to the approach in Chapters IV and V, where we restricted our initial data (to the solution spaces) so as to have well-posedness, that is, the restriction of A generated a strongly continuous semigroup.

When A generates a C-regularized semigroup, then we may construct a Frechet space, W, such that X is continuously embedded between $C(W)$ and W, on which an extension of A generates a strongly continuous semigroup. When the regularized semigroup is exponentially bounded, then this enlarged space may be chosen to be a Banach space.

This leads to characterizations of generators of regularized semigroups, in terms of generators of strongly continuous semigroups, on spaces larger than X and smaller than X (Theorems 6.6 and 6.7).

Throughout this chapter, C will be a bounded, injective operator with dense range such that $CA \subseteq AC$.

Theorem 6.1. *Suppose A generates a strongly uniformly continuous bounded C-regularized semigroup $\{W(t)\}_{t\geq 0}$. Then there exists a Banach space W and an operator B such that*

(1) *C extends to a bounded operator, $\tilde{C}$, on W,*

(2) *B generates a strongly continuous contraction semigroup on W, such that $e^{tB}\tilde{C} = \tilde{C}e^{tB}$, for all $t \geq 0$,*

(3) *$Ce^{tB}x = W(t)x$, for all $t \geq 0, x \in X$,*

(4) *$A = B|_X$ and*

(5) *$[\tilde{C}(W)] \hookrightarrow X \hookrightarrow W$.*

W may be chosen so that

$$\|x\|_W = \sup\{\|W(t)x\| \,|\, t \geq 0\}, \forall x \in X.$$

Theorem 6.2. *Suppose A generates a C-regularized semigroup $\{W(t)\}_{t\geq 0}$. Then there exists a Frechet space W and an operator B such that*

(1) *C extends to a bounded operator, $\tilde{C}$, on W,*

(2) B generates a locally equicontinuous semigroup on W, such that $e^{tB}\tilde{C} = \tilde{C}e^{tB}$, for all $t \geq 0$,

(3) $\tilde{C}e^{tB}x = W(t)x$, for all $t \geq 0, x \in X$,

(4) $A = B|_X$ and

(5) $[\tilde{C}(W)] \hookrightarrow X \hookrightarrow W$.

W may be chosen to have topology generated by the seminorms

$$\|x\|_{a,b} = \sup\{\|W(t)x\| \mid t \in [a,b]\}, \forall x \in X, a, b \in Q^+.$$

Example 6.3. Let A be as in Example 3.2,

$$(Af)(x) \equiv xf(x),$$

with maximal domain, on $C_0(\mathbf{R})$. We showed that A generates a C-regularized semigroup, where $(Cf)(x) \equiv e^{-x^2}f(x)$. Then Y, from the proof of Theorem 6.2, the completion of $C(X)$, equals the set of all continuous $f : \mathbf{R} \to \mathbf{C}$ such that

$$\|f\|_Y \equiv \sup_{x \in \mathbf{R}} |f(x)e^{-x^2}| < \infty.$$

Thus W, from Theorem 6.2, equals the set of all continuous f such that

$$\|f\|_{a,b} \equiv \sup_{x \in \mathbf{R}, t \in [a,b]} |f(x)e^{tx}e^{-x^2}| < \infty,$$

for all $a, b \in Q^+$, topologized by that family of seminorms.

The extension B is still multiplication by x and e^{tB} is multiplication by e^{tx}, on W.

Note that this choice of W is far from minimal. If g is any continuous nonvanishing function such that $x \mapsto e^{tx}g(x)$ is bounded on $\mathbf{R}$, for any $t \geq 0$, then we similarly choose W_g as the set of all continuous f such that

$$\|f\|_{a,b,g} \equiv \sup_{x \in \mathbf{R}, t \in [a,b]} |f(x)e^{tx}g(x)| < \infty,$$

for all $a, b \in Q^+$, topologized by that family of seminorms.

Thus our enlarged spaces are not sharp, as were the solution spaces of Chapters IV and V; compare this example with Example 4.16.

56

Proposition 6.4. *Suppose there exists a Banach space W, a bounded extension, $\tilde{C}$, of C, on W and an operator B such that*

$$[\tilde{C}(W)] \hookrightarrow X \hookrightarrow W,$$

B generates a strongly continuous contraction semigroup on W, $e^{tB}\tilde{C} = \tilde{C}e^{tB}$, for all $t \geq 0$ and $A = B|_X$. Then an extension of A, $C^{-1}AC$, generates a bounded strongly uniformly continuous C-regularized semigroup on X.

Proposition 6.5. *Suppose there exists a Frechet space W, a bounded extension, $\tilde{C}$, of C, to W and an operator B such that*

$$[\tilde{C}(W)] \hookrightarrow X \hookrightarrow W,$$

B generates a locally equicontinuous semigroup on W, $e^{tB}\tilde{C} = \tilde{C}e^{tB}$ and $A = B|_X$. Then an extension of A, $C^{-1}AC$, generates a C-regularized semigroup on X.

Theorem 6.6. *The following are equivalent, if $\rho(A)$ is nonempty.*

(a) *The operator A generates a bounded strongly uniformly continuous C-regularized semigroup.*

(b) *There exists a Banach space Y such that*

$$[C(X)] \hookrightarrow Y \hookrightarrow X,$$

and $A|_Y$ generates a strongly continuous contraction semigroup.

(c) *There exists a Banach space W and an operator B such that B generates a strongly continuous contraction semigroup on W, C extends to a bounded operator, $\tilde{C}$, on W, $e^{tB}\tilde{C} = \tilde{C}e^{tB}$, for all $t \geq 0$,*

$$[\tilde{C}(W)] \hookrightarrow X \hookrightarrow W,$$

and $A = B|_X$.

Theorem 6.7. *The following are equivalent, if $\rho(A)$ is nonempty.*

(a) *The operator A generates a C-regularized semigroup.*

(b) *There exists a Frechet space Y such that*

$$[C(X)] \hookrightarrow Y \hookrightarrow X,$$

and $A|_Y$ generates a locally equicontinuous semigroup.

(c) *There exists a Frechet space W and an operator B such that B generates a locally equicontinuous semigroup on W, C extends to a bounded operator $\tilde{C}$, on W, $e^{tB}\tilde{C} = \tilde{C}e^{tB}$, for all $t \geq 0$,*

$$[\tilde{C}(W)] \hookrightarrow X \hookrightarrow W,$$

and $A = B|_X$.

Proof of Theorem 6.1: Let Y be the completion of X, with respect to the norm $\|x\|_Y \equiv \|Cx\|$. Extend C to Y by defining $\tilde{C}y = \lim_{n\to\infty} Cx_n$, with the limit taken in X, whenever $\{x_n\}_{n=1}^{\infty}$ is a sequence in X converging to y, in Y. It is not hard to see that $\tilde{C}$ is bounded and injective on Y, and $\tilde{C}(Y)$ equals the closure, in X, of $C(X)$. For any $t \geq 0$, extend $W(t)$ to Y by $\tilde{W}(t)y \equiv \tilde{C}^{-1}W(t)Cy$; note that, since $W(t)$ is bounded and commutes with C, $W(t)Cy \in \tilde{C}(Y)$, for all $y \in Y$.

Let $\tilde{A}$ be the generator of the $\tilde{C}$-regularized semigroup $\{\tilde{W}(t)\}_{t\geq 0}$. Let W be the Hille-Yosida space for $\tilde{A}$. Note that, by Proposition 5.18,

$$\|x\|_W \equiv \sup_{t\geq 0} \|\tilde{C}^{-1}\tilde{W}(t)x\|_Y = \sup_{t\geq 0} \|W(t)x\|.$$

Thus $X \subseteq W$ and (5) is clear.

Let $B \equiv \tilde{A}$. To see that $A = B|_X$, suppose $x \in \mathcal{D}(B|_X)$. Then, for all $t > 0$,

$$\frac{1}{t}\left(W(t)x - Cx\right) = \tilde{C}\left(\frac{1}{t}(e^{tB}x - x)\right),$$

which converges to CBx, as $t \to 0$, in X, because $[\tilde{C}(W)] \hookrightarrow X$. Thus $x \in \mathcal{D}(A)$ and $Ax = Bx$, so that $B|_X \subseteq A$. Conversely, suppose $x \in \mathcal{D}(A)$. Then, for all $t \geq 0$,

$$\tilde{C}e^{tB}x - \tilde{C}x = W(t)x - \tilde{C}x = \int_0^t W(s)Ax\,ds = \tilde{C}\left(\int_0^t e^{sB}Ax\,ds\right),$$

so that, since $\tilde{C}$ is injective,

$$e^{tB}x - x = \int_0^t e^{sB}Ax\,ds,$$

which implies that $x \in \mathcal{D}(B)$ and $Bx = Ax \in X$, so that $x \in \mathcal{D}(B|_X)$, as desired. ∎

Proof of Theorem 6.2: This follows from Chapter IV exactly as Theorem 6.1 followed from Chapter V. ∎

Proof of Propositions 6.4 and 6.5: For $x \in X$, let $W(t)x \equiv \tilde{C}e^{tB}x$. Boundedness and strong uniform continuity follow from the fact that $[\tilde{C}(W)] \hookrightarrow X$. Since $\tilde{C}e^{tB} = e^{tB}\tilde{C}$, $\{W(t)\}_{t\geq 0}$ is a C-regularized semigroup.

Let $\tilde{A}$ be the generator of $\{W(t)\}_{t\geq 0}$ and suppose $x \in \mathcal{D}(A)$. Since $x \in \mathcal{D}(B)$ and $[\tilde{C}(W)] \hookrightarrow X$, $W(t)x$ is differentiable, with $\frac{d}{dt}W(t)x|_{t=0} = CBx = CAx$. Thus $x \in \mathcal{D}(\tilde{A})$, with $\tilde{A}x = Ax$, so that an extension of A generates $\{W(t)\}_{t\geq 0}$. To apply Proposition 3.11, we need to consider $x \in \mathcal{D}(\tilde{A})$. Then $W(t)Cx \in Im(C)$, with $C^{-1}W(t)Cx = W(t)x$, for all $t \geq 0$, thus $\frac{d}{dt}C^{-1}W(t)Cx|_{t=0}$ exists, in the norm of X, and equals $C\tilde{A}x$; since $[\tilde{C}(W)] \hookrightarrow X$, this implies that $\frac{d}{dt}W(t)Cx|_{t=0}$ exists, in the norm of W, and equals $C^2\tilde{A}x$. Since B is the generator of the $(\tilde{C})^2$-regularized semigroup $\{e^{tB}(\tilde{C})^2\}_{t\geq 0} = \{W(t)\tilde{C}x\}_{t\geq 0}$ (see Proposition 3.10), this implies that $Cx \in \mathcal{D}(B)$, with $Bx = \tilde{A}x$, thus $Cx \in \mathcal{D}(B|_X) = \mathcal{D}(A)$. Thus, $C(\mathcal{D}(\tilde{A})) \subseteq \mathcal{D}(A)$, so that, by Proposition 3.11, $\tilde{A} = C^{-1}AC$. ∎

Proof of Theorem 6.6: This follows from Theorems 5.17 and 6.1 and Propositions 6.4 and 3.11. ∎

Proof of Theorem 6.7: This follows from Theorems 4.15 and 6.2 and Propositions 6.5 and 3.11. ∎

VII. ENTIRE VECTORS AND ENTIRE EXISTENCE FAMILIES

This chapter treats the possibility of having entire solutions of the abstract Cauchy problem. We shall see that surprisingly many operators have such solutions, for a dense set of initial data, accessible through an entire C-regularized group, with the image of C dense.

One of the functional calculus constructions of e^{tA} mentioned in Chapter I is the power series representation for the exponential function. For certain vectors x, known as *entire vectors for A* (Definition 7.5), this will define an entire vector-valued function, which will be an entire solution of the abstract Cauchy problem (0.1). The existence of an entire C-existence family will be seen to correspond to the image of C being contained in the set of entire vectors. This will provide a simple way to determine for what C an operator A has an entire C-existence family.

We should remark here that a strongly continuous semigroup is entire only in the essentially trivial case where the generator is bounded (see Remark 7.11).

Definition 7.1. If $\{W(z)\}_{z \in \mathbf{C}}$ is an entire family of bounded operators, that is, the map $z \mapsto W(z)$, from $\mathbf{C}$ into $B(X)$, is entire, such that, for all $\theta \in \mathbf{R}$, the family $\{W(te^{i\theta})\}_{t \geq 0}$ is a mild C-existence family, we will say that $\{W(z)\}_{z \in \mathbf{C}}$ is an *entire C-existence family*.

Note that Proposition 2.7 implies that for all $x \in X$, $W(z)x \in \mathcal{D}(A)$ and $\frac{d}{dz}W(z)x = AW(z)x$, for all $z \in \mathbf{C}$. In particular, $\{W(te^{i\theta})\}_{t \geq 0}$ is a strong C-existence family, for all $\theta \in \mathbf{R}$.

Definition 7.2. A strongly continuous family of bounded operators $\{W(t)\}_{t \in \mathbf{R}}$ is a *C-regularized group* if $W(0) = C$ and $W(t)W(s) = CW(t+s)$, for all real s, t.

The *generator* is defined exactly as with C-regularized semigroups, except that in "$\lim_{t \to 0}$," t may be positive or negative.

Essentially the same argument as with strongly continuous groups (see [Paz, Section 1.6]) shows that A generates a C-regularized group if and only if both A and $(-A)$ generate C-regularized semigroups.

The following is *not* a natural definition to make, when dealing with strongly continuous semigroups (see Remark 7.11).

Definition 7.3. An *entire C-regularized group* is an entire family of bounded operators $\{W(z)\}_{z \in \mathbf{C}}$ such that

$$W(0) = C, \quad CW(z + w) = W(z)W(w) \quad (z, w \in \mathbf{C}).$$

60

Example 7.4. Let A be as in Example 3.2,

$$(Af)(x) \equiv xf(x),$$

with maximal domain, on $C_0(\mathbf{R})$. Then it is clear how $\{W(t)\}_{t \geq 0}$ extends to an entire C-regularized group

$$(W(z)f)(x) \equiv e^{-x^2} e^{zx} f(x), \quad \text{for } x \text{ real, } z \text{ complex.}$$

Definition 7.5. Let $C^{\infty}(A) \equiv \bigcap_{n=0}^{\infty} \mathcal{D}(A^n)$. The C^{∞} vector x is an *entire vector for A* if

$$\sum_{n=0}^{\infty} \frac{s^n}{n!} \|A^n x\| < \infty,$$

for all $s > 0$.

We will write $\mathcal{E}(A)$ for the set of all entire vectors for A.
For $x \in \mathcal{E}(A), z \in \mathbf{C}$, we define

$$e^{zA} x \equiv \sum_{k=0}^{\infty} \frac{z^k}{k!} A^k x.$$

It is clear that, for $x \in \mathcal{E}(A)$, this defines an entire family of vectors $\{e^{zA}x\}_{z \in \mathbf{C}}$.

Remark 7.6. It is well known that a generator of a bounded, strongly continuous group has a core of entire vectors. This may be shown in the following manner. If iA is the generator, then $-A^2$ generates a bounded strongly continuous holomorphic semigroup given by

$$e^{-zA^2} x = (4\pi z)^{-\frac{1}{2}} \int_{\mathbf{R}} (e^{itA} x) e^{\frac{-t^2}{4z}} \, dt,$$

for $x \in X, Re(z) > 0$ (this is just the Fourier inversion formula for the function $s \mapsto e^{-zs^2}$; see Chapter XII).

Then it is not hard to verify that $Im(e^{-zA^2}) \subseteq \mathcal{E}(A)$, whenever $Re(z) > 0$; thus, since

$$x = \lim_{n \to \infty} e^{-\frac{1}{n}A^2} x \ (x \in X), \quad \text{and} \quad Ax = \lim_{n \to \infty} e^{-\frac{1}{n}A^2} Ax \ (x \in \mathcal{D}(A),$$

$\mathcal{E}(A)$ is a core.

The proof of the following consists of straightforward series arguments.

Proposition 7.7.

(1) $x \in \mathcal{E}(A)$ if and only if $x \in C^\infty(A)$ and

$$\sum_{k=0}^{N} \frac{z^k}{k!} A^k x$$

converges uniformly in z on compact subsets of $\mathbf{C}$, as $N \to \infty$.

(2) The operator A maps $\mathcal{E}(A)$ into itself, with $e^{zA}Ax = Ae^{zA}x = \frac{d}{dz}e^{zA}x$, for all $x \in \mathcal{E}(A), z \in \mathbf{C}$.

(3) For all complex z, e^{zA} maps $\mathcal{E}(A)$ into itself, with $e^{zA}e^{wA}x = e^{(z+w)A}x$, for all $x \in \mathcal{E}(A), w \in \mathbf{C}$.

(4) For any $\theta \geq 0, \mathcal{E}(A)$ is contained in the solution space for $e^{i\theta}A$, with

$$u(t,x) = e^{te^{i\theta}A}x,$$

for all $x \in \mathcal{E}(A)$.

Perhaps the most natural way to think of $\mathcal{E}(A)$ is as a subspace of Z, the solution space (see Chapter IV), consisting of the solutions of the abstract Cauchy problem with entire extensions.

Theorem 7.8. *Suppose A is closed. Then the following are equivalent.*

(a) $Im(C) \subseteq \mathcal{E}(A)$.

(b) There exists an entire C-existence family for A.

The entire C-existence family for A is then given by

$$W(z) \equiv e^{zA}C,$$

where e^{zA} is as in Definition 7.5.

Corollary 7.9. *Suppose $\{W(z)\}_{z \in \mathbf{C}}$ is an entire C-existence family for A. Then for all $z \in \mathbf{C}$,*

(1) for all $k \in \mathbf{N}, (\frac{d}{dz})^k W(z) = A^k W(z) \in B(X)$; and

(2) $W(z)$ maps X into $\mathcal{E}(A)$.

Corollary 7.10. *Suppose $CA \subseteq AC$ and A is closed. Then the following are equivalent.*

(a) $Im(C) \subseteq \mathcal{E}(A)$.

(b) An extension of A generates an entire C-regularized group that maps X into $\mathcal{E}(A)$.

Remark 7.11. The definition analogous to Definition 7.3, for strongly continuous semigroups (or, more generally, for exponentially bounded n-times integrated semigroups—see Chapter XVIII), reduces to the trivial case of the generator being bounded. In particular, if $\pm A$ and $\pm iA$ all generate strongly continuous semigroups (or exponentially bounded n-times integrated semigroups), then A must be bounded. This may be easily shown, using the following fact.

Proposition. *Suppose A is a closed operator whose spectrum is contained in a bounded set, with $\|(w - A)^{-1}\|$ $O(|w|^k)$, for some integer k, for w large. Then $A \in B(X)$.*

Proof of Theorem 7.8: (a) $\rightarrow$ (b). It is clear from Proposition 7.7 and Theorem 4.13 that

$$W(z)x \equiv e^{zA}Cx \ (x \in X)$$

is the desired entire C-existence family for A.

(b) $\rightarrow$ (a). We will show the following by induction on n. For all $n \in \mathbf{N}, z \in \mathbf{C}, x \in X$,

$$W(z)x \in \mathcal{D}(A^n), \text{ and } (\frac{d}{dz})^n W(z)x = A^n W(z)x. \tag{7.12}$$

By Proposition 2.7, (7.12) is true for $n = 1$. Suppose (7.12) is true for arbitrary $n \in \mathbf{N}, z \in \mathbf{C}, x \in X$.

Let $x_k \equiv k((\frac{d}{dz})^{n-1}W(z+\frac{1}{k})x - (\frac{d}{dz})^{n-1}W(z)x)$, then $x_k \rightarrow (\frac{d}{dz})^n W(z)x$ as $k \rightarrow \infty$, and by the induction hypothesis,

$$Ax_k = k(A^n W(z+\frac{1}{k})x - A^n W(z)x) = k((\frac{d}{dz})^n W(z+\frac{1}{k})x - (\frac{d}{dz})^n W(z)x),$$

which converges to $(\frac{d}{dz})^{n+1}W(z)x$, as $k \rightarrow \infty$, so that, since A is closed, $A^n W(z)x = (\frac{d}{dz})^n W(z)x \in \mathcal{D}(A)$, with $A^{n+1}W(z)x = (\frac{d}{dz})^{n+1}W(z)x$, completing the induction and proving (7.12).

The assertion (7.12) now implies that $x \in C^\infty(A)$, and the MacLaurin series for $W(z)$ is given by

$$W(z) = \sum_{k=0}^{\infty} \frac{z^k}{k!} A^k,$$

with the convergence of the sum uniform in z, on compact subsets of $\mathbf{N}$. By Proposition 7.7(1), $Im(C) \subseteq \mathcal{E}(A)$, as desired. ∎

Proof of Corollary 7.10: (a) → (b). A calculation shows that $W(z) \equiv e^{zA}C$, as in Theorem 7.8, is an entire C-regularized group. Since $CA \subseteq AC, C$ leaves $\mathcal{E}(A)$ invariant, with $e^{zA}Cx = Ce^{zA}x$, for all $z \in \mathbf{C}$. Proposition 7.7 now implies that $W(z)$ maps X into $\mathcal{E}(A)$, for all z. By Theorem 3.7(a), an extension of A generates $\{W(z)\}_{z \in \mathbf{C}}$.

(b) → (a). Since A is closed, Theorem 3.8(1) implies that $\{W(z)\}_{z \in \mathbf{C}}$ is an entire C-existence family for A. Thus (a) follows from Theorem 7.8. ∎

Proof of Remark 7.11: The Proposition follows by the construction of the Riesz-Dunford functional calculus, except that we do not yet know that A is bounded, so we cannot draw any conclusions from the properties of the Riesz-Dunford functional calculus.

Define $T \in B(X)$ by

$$T \equiv \int_{|z|=M} (w - A)^{-1} \frac{dw}{2\pi i}$$

for M sufficiently large. We want to show that $T = I$, the identity operator. For this, note that, by the residue theorem,

$$2\pi i (T - I) = \int_{|z|=M} \left((w - A)^{-1} - \frac{1}{w} \right) dw$$

$$= \int_{|z|=M} ((w - A + A)(w - A)^{-1} - 1) \frac{dw}{w} \qquad (7.13)$$

$$= \int_{|z|=M} A(w - A)^{-1} \frac{dw}{w}.$$

Fix $\omega \in \rho(A)$. We will show by induction that, for any $n \in \mathbf{N}$,

$$2\pi i (T - I)(A - \omega)^{-n} = \int_{|z|=M} A^n (A - \omega)^{-n} (w - A)^{-1} \frac{dw}{w^n}. \qquad (7.14)$$

For $n = 1$, (7.14) follows by applying $(A - \omega)^{-1}$ to both sides of (7.13). Suppose (7.14) holds for $n = m \geq 1$. Then

$$2\pi i (T - I)(A - \omega)^{-(m+1)}$$

$$= \int_{|z|=M} A^m (A - \omega)^{-(m+1)} (w - A)^{-1} \frac{dw}{w^m}$$

$$= \int_{|z|=M} (w - A + A) A^m (A - \omega)^{-(m+1)} (w - A)^{-1} \frac{dw}{w^{m+1}}$$

$$= - \int_{|z|=M} A^m (A - \omega)^{-(m+1)} \frac{dw}{w^{m+1}}$$

$$+ \int_{|z|=M} A^{m+1} (A - \omega)^{-(m+1)} (w - A)^{-1} \frac{dw}{w^{m+1}};$$

by the Cauchy integral formula, the first integral is zero, thus this completes the induction, proving (7.14).

Choosing $n > k + 1$ and letting $M \to \infty$ in (7.14) now implies that

$$(A - \omega)^{-n}(T - I) = (T - I)(A - \omega)^{-n} = 0,$$

so that, by applying $(A - \omega)^n$ to both sides, we conclude that $T = I$.
Since A is closed, $X = Im(T) \subseteq \mathcal{D}(A)$, with

$$Ax = ATx = \int_{|z|=M} A(w - A)^{-1}x \, \frac{dw}{2\pi i} \quad \forall x \in X.$$

This implies that $A \in B(X)$. ∎

65

VIII. REVERSIBILITY OF PARABOLIC PROBLEMS

Imagine a world where diffusion is reversible. A consequence of the results in this chapter is the existence of such a Frechet space, that is, a space on which the Laplacian generates a strongly continuous *group* (see Example 8.6). More generally, we use regularized semigroups to show that any parabolic problem is reversible on a dense set, with well-posedness on a dense Frechet subspace, as guaranteed by Theorem 4.8 (see Corollary 8.3). More precisely, we show that, if $-A$ generates a strongly continuous holomorphic semigroup, then A generates an entire C-existence family, with $Im(C)$ dense (Theorem 8.2).

We will need fractional powers. Since the complex-valued function $z \mapsto z^\alpha$ is analytic on $\mathbf{C} - (-\infty, 0]$, to define fractional powers of an operator B, we need to have B contained in $\mathbf{C} - (-\infty, 0]$, in some sense.

Definition 8.1. Suppose B is a closed, densely defined operator such that $(-\infty, 0)$ is contained in $\rho(B)$, with $\{r \| (r + B)^{-1} \| \mid r > 0\}$ bounded.

It can be shown (see [Balak] or [F1]) that there exists $\theta < \pi$ such that the spectrum of B is contained in $\overline{S_\theta} = \{re^{i\phi} \mid r \geq 0, |\phi| < \theta\}$ and

$$\{\| z(z - B)^{-1} \| \mid z \notin S_\theta\}$$

is bounded.

Suppose first that zero is in $\rho(B)$. For $0 < \alpha\theta < \frac{\pi}{2}$, we then define the *fractional power* $(-B^\alpha)$, as the generator of the exponentially decaying strongly continuous holomorphic semigroup $\{T_\alpha(t)\}_{t \geq 0}$, defined by

$$T_\alpha(t) = \int_{\Gamma_\phi} e^{-tw^\alpha} (w - A)^{-1} \frac{dw}{2\pi i},$$

where $\pi > \phi > \theta$, $\alpha\phi < \frac{\pi}{2}$, $t > 0$, Γ_ϕ = boundary of S_ϕ (see [F1]).

When zero is not in $\rho(B)$, another formula (also in [F1]) defines the fractional powers, with the same properties. We will not need this formula, merely the existence of fractional powers of operators as in the above definition.

Suppose $(-B)$ generates a strongly continuous holomorphic semigroup. Then there exists real k, and $\theta < \frac{\pi}{2}$, such that the spectrum of $(B - k)$ is contained in S_θ, with the growth condition on the resolvent of Definition 8.1. If $\alpha > 1$ is chosen so that $\alpha\theta < \frac{\pi}{2}$, then the function $w \mapsto e^{-w^\alpha} e^{zw}$ is bounded on S_θ, for any complex z. Thus, in the following theorem, we expect $(B - k)$, and hence B, to generate an entire $e^{-(B-k)^\alpha}$-regularized group, which we think of as $e^{-(B-k)^\alpha} e^{zB}$. We could construct this regularized group with the Cauchy integral formula, with a construction similar to that which was used in Definition 8.1 (see Remark 8.5); however, it is simpler to use entire vectors, as in the previous Chapter, to show that such a group exists.

Theorem 8.2. *Suppose $-B$ generates a strongly continuous holomorphic semigroup of angle θ. Then there exists $k \in \mathbf{R}$ such that B generates an entire $e^{-(B-k)^\alpha}$-regularized group, whenever $1 < \alpha < \frac{\frac{\pi}{2}}{\frac{\pi}{2}-\theta}$.*

Theorem 4.15 then gives us the following.

Corollary 8.3. *Suppose $-B$ generates a strongly continuous holomorphic semigroup. Then*

(1) *the abstract Cauchy problem (0.1) has a unique entire solution, for all initial data x in a dense subspace; and*

(2) *there exists Z, dense in X, such that Z is a Frechet space,*

$$[Im(e^{-(B-k)^\alpha})] \hookrightarrow Z \hookrightarrow X,$$

where k and α are as in Theorem 8.2, and $B|_Z$ generates an entire strongly continuous group.

Example 8.4. Unlike the generator of a strongly continuous holomorphic semigroup, the generator of a strongly continuous semigroup may not generate a C-regularized group, for any choice of C. Let $X \equiv \{f \in C[0,1] \mid f(0) = 0\}$, and B equal the generator of the strongly continuous semigroup $\{e^{tB}\}_{t\geq 0}$ on X, defined by right translation,

$$(e^{tB}f)(x) \equiv \begin{cases} f(x-t) & x \geq t \\ 0 & x \leq t \end{cases}.$$

If B did generate a C-regularized group, $\{W(t)\}_{t\in\mathbf{R}}$, then by the uniqueness of the solutions of the abstract Cauchy problem (see Theorem 3.5), $W(t)$ would equal Ce^{tB}, for $t \geq 0$.

Note that, for all real t, $W(t)$ must be injective; for, if $W(t)x = 0$, then $C^2x = W(-t)W(t)x = 0$, so that, since C is injective, $x = 0$.

For $t > 1$, $e^{tB} \equiv 0$. Thus Ce^{tB} is not injective, for any choice of C, so that B cannot generate a C-group.

More generally, this argument shows that, if B generates a strongly continuous semigroup $\{e^{tB}\}_{t\geq 0}$, and B generates a C-regularized group, then e^{tB} is injective, for all $t \geq 0$.

67

Remark 8.5. It may be shown that the regularized group of Theorem 8.2 is given by

$$W(z) = \int_{\Gamma_\phi} e^{zw} e^{-w^\alpha} (w + k - B)^{-1} \frac{dw}{2\pi i},$$

which we think of as $e^{zA} e^{-A^\alpha}$, as with the Riesz-Dunford functional calculus (see Chapter XXII).

Example 8.6: Backward Heat Equation. The *backward heat equation* has the form

$$\frac{\partial u}{\partial t}(\vec{x}, t) + \Delta u(\vec{x}, t) = 0 \quad (\vec{x} \in D, \; t \geq 0)$$

$$u(\vec{x}, t) = 0 \quad (\vec{x} \in \partial D, \; t \geq 0)$$

$$u(\vec{x}, 0) = f(\vec{x}) \quad (\vec{x} \in D),$$

where D is a bounded open set in $\mathbf{R}^n$ with smooth boundary ∂D.

Theorem 8.2 yields unique global solutions in $X \equiv L^p(D)$ $(1 \leq p < \infty)$; in fact, the map $t \mapsto u(\cdot, t)$, from $[0, \infty)$ into X, extends to an entire function, for f in a dense subspace of X, because $A \equiv \Delta$, $\mathcal{D}(A) \equiv W^{2,p}(D) \cap W_0^{1,p}(D)$, generates a strongly continuous holomorphic semigroup on those spaces.

This is saying that the heat equation is reversible on a dense set. What is most surprising is that this produces a Frechet space, Z, dense and continuously embedded in $L^p(D)$, on which this reversibility is well-posed, in the strongest sense; that is, $-\Delta$ generates a strongly continuous semigroup (see Theorems 4.8 and 4.13). This is hard to imagine, a space where diffusion is continuously reversible.

Lemma 8.7. *Suppose $\sigma(A) \subseteq \Omega$, $\partial\Omega$ is a finite union of smooth (possibly unbounded) curves, $g : \partial\Omega \to \mathbf{C}$,*

(1)
$$\int_{\partial\Omega} g(z)\, dz = 0,$$

  and

(2)
$$\int_{\partial\Omega} |zg(z)| \|(z - A)^{-1}x\| \, d|z|, \qquad \int_{\partial\Omega} |g(z)| \|A(z - A)^{-1}x\| \, d|z|,$$

  and
$$\int_{\partial\Omega} |g(z)| \|(z - A)^{-1}x\| \, d|z|$$

  are finite.

Then $y \equiv \int_{\partial \Omega} g(z)(z-A)^{-1}x\,dz \in \mathcal{D}(A)$, with

$$Ay = \int_{\partial \Omega} zg(z)(z-A)^{-1}x\,dz.$$

Proof: This follows from the fact that A is closed, and, by writing $A = z - (z - A)$,

$$\int_{\partial \Omega} A\left(g(z)(z-A)^{-1}x\right)\,dz = \int_{\partial \Omega} zg(z)(z-A)^{-1}x\,dz - \left(\int_{\partial \Omega} g(z)\,dz\right)x$$

$$= \int_{\partial \Omega} zg(z)(z-A)^{-1}x\,dz.$$

■

Lemma 8.8. *Suppose $\{T(z)\}_{z \in S_\theta}$ is a strongly continuous holomorphic semigroup. Then $T(z)$ is injective and has dense range, for all $z \in S_\theta$.*

Proof: Suppose $z_0 \in S_\theta$ and $T(z_0)x = 0$. Then for all $n \in \mathbf{N}$,

$$(\frac{d}{dz})^n T(z_0)x = A^n T(z_0)x = 0,$$

thus, since $T(z)$ is holomorphic in S_θ, $T(z)x = 0$, for all $z \in S_\theta$, so that $x = \lim_{t \to 0} T(t)x = 0$.

This shows that $T(z_0)$ is injective. To show that it has dense range, suppose x^* is in the annihilator of the image of $T(z_0)$. Then

$$0 = <x^*, T(z_0)x> = <T(z_0)^*x^*, x>,$$

for all $x \in X$, thus

$$0 = (A^*)^n T(z_0)^* x^* = (\frac{d}{dz})^n T(z_0)^* x^*,$$

for all $n \in \mathbf{N}$. As in the previous paragraph, this implies that

$$0 = <x^*, T(z)x>, \forall z \in S_\theta,$$

so that $<x^*, x> = 0$, for all $x \in X$. Thus $T(z_0)$ has dense range. ■

Proof of Theorem 8.2: There exist positive θ and real k, such that $(k - B)$ generates an exponentially decaying strongly continuous holomorphic semigroup of angle θ. This means that we may define fractional powers of $(B - k)$, as in Definition 8.1.

Let

$$C \equiv e^{-(B-k)^\alpha} \equiv \int_{\Gamma_\phi} e^{-w^\alpha}(w+k-B)^{-1}\frac{dw}{2\pi i},$$

where $\frac{\pi}{2} > \phi > (\frac{\pi}{2} - \theta), \phi < \alpha\phi < \frac{\pi}{2}$ and Γ_ϕ is as in Definition 8.1.

We will show that $Im(C) \subseteq \mathcal{E}(A)$ (see Definition 7.5).

By Lemma 8.7 and the residue theorem, with $g(w) \equiv w^{n-1}e^{-w^\alpha}$, $Im(C) \subseteq C^\infty(A)$, with

$$A^n Cx = \int_{\Gamma_\phi} w^n e^{-w^\alpha}(w+k-B)^{-1}x\,\frac{dw}{2\pi i},$$

for any $x \in X, n+1 \in \mathbf{N}$; thus, for any $t \geq 0$,

$$\sum_{n=0}^\infty \frac{t^n}{n!}\|A^n Cx\| \leq \int_{\Gamma_\phi} e^{t|w|}|e^{-w^\alpha}|\,\|(w+k-B)^{-1}x\|\,\frac{d|w|}{2\pi},$$

which is finite.

This shows that the image of C is contained in the set of entire vectors for B. By Corollary 7.10, an extension of B generates an entire C-regularized group. Since $\rho(B)$ is nonempty, it is generated by B (see Proposition 3.9). ∎

70

IX. THE CAUCHY PROBLEM FOR THE LAPLACE EQUATION

We have already seen how regularized semigroups cover extreme cases of ill-posedness, such as the backwards heat equation. Another famous example of an ill-posed problem is the *Cauchy problem for the Laplace equation in an infinite cylinder*

$$\Delta u(x, \vec{y}) = 0 \quad ((x, \vec{y}) \in [0, \infty) \times D)$$

$$u(x, \vec{y}) = 0 \quad ((x, \vec{y}) \in [0, \infty) \times \partial D)$$

$$u(0, \vec{y}) = f(\vec{y}) \quad (\vec{y} \in D) \tag{9.1}$$

$$\frac{\partial u}{\partial x}(0, \vec{y}) = g(\vec{y}) \quad (\vec{y} \in D),$$

where D is a bounded open set in $\mathbf{R}^n$, with smooth boundary. This may, as with the backward heat equation (see Chapter VIII), be shown to have unique entire solutions, for all f, g in a dense subspace of $X \equiv L^p(D)$ $(1 \leq p < \infty)$. As with Corollary 8.3, we will also, by Theorems 4.13 and 3.8(3), have well-posedness on a Frechet space densely and continuously embedded in $L^p(D)$.

This will follow by writing (9.1) as

$$(\frac{d}{dx})^2 u(x) = Bu(x), (x \geq 0), u(0) = f, (\frac{d}{dx})u(0) = g,$$

where $B \equiv -\Delta$, on $L^p(D)$, with $\mathcal{D}(B) \equiv W^{2,p}(D) \cap W_0^{1,p}(D)$. The relevant property of B is that $-B$ generates a bounded strongly continuous semigroup.

With the usual matrix reduction, this becomes a first-order abstract Cauchy problem,

$$\frac{d}{dx}\vec{u}(x) = \mathcal{A}\vec{u}(x), (x \geq 0) \ \vec{u}(0) = (f, g), \tag{9.2}$$

where

$$\mathcal{A} \equiv \begin{bmatrix} 0 & I \\ B & 0 \end{bmatrix}.$$

In this chapter, we will show that there exists an operator C, with dense range, such that an extension of $\mathcal{A}$ generates an entire C-regularized group. More generally, we will show the following.

71

Theorem 9.3. *Suppose there exists complex λ such that $(-\infty, 0) \subseteq \rho(\lambda B)$ and $\{r\|(r + \lambda B)^{-1}\| \mid r > 0\}$ is bounded. Then there exists $\alpha > \frac{1}{2}$ such that an extension of $\mathcal{A}$, on $X \times X$, $\mathcal{D}(\mathcal{A}) \equiv \mathcal{D}(B) \times X$, generates an entire $e^{-(\lambda B)^\alpha}$-regularized group that leaves $\mathcal{D}(\mathcal{A})$ invariant.*

Corollary 9.4. *Suppose B is as in Theorem 9.3. Then*

(1) *(9.2) has a unique entire solution, for all initial data (f, g) in a dense subspace; and*

(2) *there exists Z, dense in $X \times X$, such that Z is a Frechet space,*

$$[Im(e^{-(\lambda B)^\alpha})] \hookrightarrow Z \hookrightarrow X \times X,$$

where α is as in Theorem 9.3, and $\mathcal{A}|_Z$ generates an entire strongly continuous group.

Proof of Theorem 9.3: There exists a square root, $G \equiv (\lambda B)^{\frac{1}{2}}$, such that $-G$ generates a bounded holomorphic strongly continuous semigroup. By Corollary 7.10 and Theorem 8.2, it is sufficient to show that, whenever x_1 and x_2 are entire vectors for G, then $\vec{x} \equiv (x_1, x_2)$ is an entire vector for $\mathcal{A}$. But this is not hard to see, when one notes that

$$(\mathcal{A})^{2n} = B^n I,$$

so that, for $t \geq 0$,

$$\sum_{k=0}^{\infty} \|\mathcal{A}^k \vec{x}\| \frac{t^k}{k!}$$

$$= \sum_{n=0}^{\infty} \left[\frac{t^{2n}}{(2n)!} \|(B^n x_1, B^n x_2)\| + \frac{t^{(2n+1)}}{(2n+1)!} \|(B^n x_2, B^{n+1} x_1)\| \right],$$

which is finite, since B maps $\mathcal{E}(G)$ into $\mathcal{E}(G)$ (see Proposition 7.7). ∎

X. BOUNDARY VALUES OF HOLOMORPHIC
SEMIGROUPS

The heat semigroup $e^{z\Delta}$, where Δ is the Laplacian on $L^p(\mathbf{R}^n)$, constitutes a holomorphic function on the right half-plane (RHP) which is bounded on every sector $S_\theta \equiv \{z \,|\, |\arg(z)| < \theta\}$, for θ less than $\frac{\pi}{2}$. Its boundary values, $e^{is\Delta}$, however, constitute a strongly continuous group (the Schrödinger group) only when $p = 2$. It is important to "regularize" $e^{is\Delta}$ for $p \neq 2$ and study the analogous Schrödinger "group" in this case.

In general, when e^{zA} is a bounded (in sectors S_θ, for $\theta < \frac{\pi}{2}$) holomorphic strongly continuous semigroup of angle $\frac{\pi}{2}$, it is natural to ask when boundary values exist, in some sense. As indicated in the first paragraph, to include many interesting examples, this "sense" needs to be weaker than the sense of a strongly continuous group.

In this chapter, we show that, when A generates a bounded strongly continuous holomorphic semigroup of angle $\frac{\pi}{2}$, then the amount of regularizing, C, required to make e^{isA} into a C-regularized group $e^{isA}C$ depends on how rapidly $\|e^{zA}\|$ grows as z approaches the imaginary axis. This is useful for applications, because we may restrict our attention to *bounded* operators e^{zA}, for $Re(z)$ strictly greater than zero, which are much easier to work with than the unbounded operators e^{isA}, whose definition, in general, may be somewhat mysterious.

More specifically, we may look at the behaviour of $\|e^{zA}\|$ in sectors S_θ, as θ approaches $\frac{\pi}{2}$, or in half-planes $Re(z) > a > 0$, as a approaches 0. By definition of a bounded holomorphic semigroup, $\|e^{zA}\|$ is bounded in S_θ, for any $\theta < \frac{\pi}{2}$; we show that iA generating a $(1 - A)^{-r}$-regularized group that is $O(1 + |s|^r)$ corresponds to $\|e^{zA}\|$ being $O((\frac{1}{\cos\theta})^r)$, where $\theta = \arg(z)$. Generating a bounded $(1 - A)^{-r}$-regularized group corresponds to $\|e^{zA}\|$ being $O((\frac{1}{Re(z)})^r)$.

We assume throughout this chapter that A generates a bounded strongly continuous holomorphic semigroup of angle $\frac{\pi}{2}$ $\{e^{zA}\}_{z \in RHP}$.

Theorem 10.1. *Suppose that $\gamma \geq 0$ and there exists $M < \infty$ such that*

$$\|e^{zA}\| \leq M\left(\frac{|z|}{Re(z)}\right)^\gamma.$$

Then for all $r > \gamma$, there exists $M_{r,\gamma} < \infty$ such that iA generates a $(1 - A)^{-r}$-regularized group $\{W_r(s)\}_{s \in \mathbf{R}}$ such that

$$\|W_r(s)\| \leq M_{r,\gamma}(1 + |s|^\gamma).$$

Theorem 10.2. *Suppose $\gamma \geq 0$. Then the following are equivalent.*

(a) *For all $r > \gamma$, there exists $M_r < \infty$ such that*

$$\|e^{zA}\| \leq M_r \left(\frac{1 + |z|}{Re(z)}\right)^r.$$

(b) *For all $r > \gamma$, there exists C_r such that iA generates a $(1 - A)^{-r}$-regularized group $\{W_r(s)\}_{s \in \mathbf{R}}$ such that*

$$\|W_r(s)\| \leq C_r(1 + |s|^r).$$

(c) *$\|(1 + |z|^r)^{-1} e^{zA}(1 - A)^{-r}\|$ is uniformly bounded in the right half-plane, for all $r > \gamma$.*

It is theoretically interesting that a *bounded* $(1 - A)^{-r}$-regularized semigroup corresponds to the same growth conditions on right half-planes $Re(z) > a$ rather than sectors. The proof of the following theorem is essentially the same as the proof of Theorem 10.2.

Theorem 10.3. *Suppose $\gamma \geq 0$. Then the following are equivalent.*

(a) *For all $r > \gamma$, there exists $M_r < \infty$ such that*

$$\|e^{zA}\| \leq M_r (Re(z))^{-r}.$$

(b) *iA generates a bounded $(1 - A)^{-r}$-regularized group, for all $r > \gamma$.*

(c) *$\|e^{zA}(1 - A)^{-r}\|$ is uniformly bounded in the right half-plane, for all $r > \gamma$.*

Proof of Theorem 10.1: Fix $r > \gamma$. We have

$$(1 - A)^{-r}x = \frac{1}{\Gamma(r)} \int_0^\infty e^{-u} u^{r-1} e^{uA} x \, du$$

([Komu, Proposition 11.1]); thus, for $z = t + is$,

$$(1 - A)^{-r} e^{zA} x = \frac{1}{\Gamma(r)} \int_0^\infty e^{-u} u^{r-1} e^{((t+u)+is)A} x \, du,$$

so that we may estimate as follows:

$$\|(1 - A)^{-r} e^{zA}\| \leq \frac{M}{\Gamma(r)} \int_0^\infty e^{-u} u^{r-1} \left(\frac{1}{u+t}\right)^\gamma (\sqrt{(u+t)^2 + s^2})^\gamma \, du$$

$$\leq \frac{M}{\Gamma(r)} \int_0^\infty e^{-u} u^{r-1-\gamma} (\sqrt{(u+t)^2 + s^2})^\gamma \, du,$$

$$(10.4)$$

74

which is convergent since $r - 1 - \gamma > -1$.

As with strongly continuous holomorphic semigroups, since $(1-A)^{-r}e^{zA}$ is holomorphic and bounded in every rectangle $\{t + is \mid 0 < t < 1, |s| < a\}, a > 0$, its boundary values exist when $t \to 0$ and define a $(1 - A)^{-r}$-regularized group, $\{W_r(s)\}_{s \in \mathbf{R}}$. It is straightforward to verify that the generator of the $(1 - A)^{-r}$-regularized semigroup $\{(1 - A)^{-r}e^{zA}\}_{z \in RHP}$ is A, thus iA generates $\{W_r(s)\}_{s \in \mathbf{R}}$.

All that remains is the growth estimate on $\|W_r(s)\| = \lim_{t \to 0} \|(1 - A)^{-r}e^{(t+is)A}\|$, which, by (10.4), is less than or equal to

$$\frac{M}{\Gamma(r)} \int_0^\infty e^{-u} u^{r-1-\gamma}(\sqrt{u^2 + s^2})^\gamma \, du.$$

The integral may be shown to be less than or equal to

$$(1 + |s|^\gamma) \int_0^\infty e^{-u} u^{r-1-\gamma}(\sqrt{1 + u^2})^\gamma \, du,$$

by considering separately $|s| \leq 1$ and $|s| \geq 1$, for if $|s| \leq 1$, then $u^2 + s^2 \leq u^2 + 1$, while if $|s| \geq 1$, then $u^2 + s^2 = s^2((\frac{u}{s})^2 + 1) \leq s^2(u^2 + 1)$. ∎

Proof of Theorem 10.2: The implications (a) $\to$ (b) and (c) $\to$ (b) are essentially the same as the proof of Theorem 10.1.

(b) $\to$ (c) is a consequence of the maximum principle for analytic functions.

(b) $\to$ (a). Fix $r > \gamma$. Since $\{e^{zA}\}$ is a bounded strongly continuous holomorphic semigroup, there exists $K_r < \infty$ such that

$$\|(1 - A)^r e^{tA}\| \leq K_r t^{-r}, \forall t > 0$$

(see [Paz, Chapter 2]).

For $z = x + iy$, with $x > 0$,

$$\|e^{zA}\| = \|(1 - A)^r e^{xA} W_r(y)\|$$

$$\leq \|(1 - A)^r e^{xA}\| \|W_r(y)\|$$

$$\leq K_r x^{-r} C_r (1 + |y|^r) \leq (K_r)(C_r)(\frac{1 + |z|}{Re(z)})^r.$$

∎

75

XI. THE SCHRÖDINGER EQUATION

A much less extreme form of ill-posedness, than appears in Chapters VIII and IX, is the Schrödinger equation, on $L^p(\mathbf{R}^n), p \neq 2$, or $C_0(\mathbf{R}^n)$,

$$\frac{\partial u}{\partial t}(t,x) = i\left(\Delta u(t,x) - V(x)u(t,x)\right) \ (t \in \mathbf{R}), \ u(0,x) = f(x).$$

We apply the previous chapter to the Schrödinger operator with potential, $i(\Delta - V)$, for V real valued, with V_+ a Kato perturbation, $V_- \in L^\infty(\mathbf{R}^n)$, and show that

$$\{e^{is(\Delta-V)}(\omega - (\Delta - V))^{-r}\}_{s \in \mathbf{R}}$$

is a polynomially bounded regularized semigroup on $L^p(\mathbf{R}^n)$, for $r > 2n|\frac{1}{p} - \frac{1}{2}|$, ω sufficiently large (Theorem 11.4). This produces polynomially bounded mild solutions of the Schrödinger equation for all initial data f in $\mathcal{D}((\Delta - V)^r)$, on $L^(\mathbf{R}^n)$.

It is well known that, even for $V \equiv 0$, the Schrödinger operator does not generate a strongly continuous semigroup on $L^p(\mathbf{R}^n), p \neq 2$, or $C_0(\mathbf{R}^n)$.

For $V \equiv 0$, we may cut r in half.

Theorem 11.1. *Suppose* $1 \le p < \infty$. *Then for all* $r > n|\frac{1}{p} - \frac{1}{2}|, i\Delta$, *on* $L^p(\mathbf{R}^n)$, *generates a* $(1-\Delta)^{-r}$-*regularized group,* $\{W_r(s)\}_{s \in \mathbf{R}}$, *that is* $O((1 + |s|^{n|\frac{1}{p}-\frac{1}{2}|}))$.

On $C_0(\mathbf{R}^n)$ *or* $BUC(\mathbf{R}^n), i\Delta$ *generates a* $(1 - \Delta)^{-r}$-*regularized group,* $\{W_r(s)\}_{s \in \mathbf{R}}$ *that is* $O(1 + |s|^{\frac{n}{2}})$, *for all* $r > \frac{n}{2}$.

Definition 11.2. We will denote by K^n the *Kato class* of measurable functions on $\mathbf{R}^n$, as defined in [Sim, p. 453]. This includes, but is not limited to, $L^\infty(\mathbf{R}^n)$.

Definition 11.3. For $V \in K^n$, it is shown in [Sim, Theorem A.2.7] that

$$H \equiv \Delta - V,$$

defined as a quadratic form, is a self-adjoint operator on $L^2(\mathbf{R}^n)$.

The following theorem states that H, with appropriate domain, generates an exponentially bounded $(\omega - H)^{-r}$-regularized group, for r twice as big as in Theorem 11.1.

76

Theorem 11.4. *Suppose* $1 \leq p < \infty$, $V_+ \in K^n$, $V_- \in L^\infty(\mathbf{R}^n)$. *Let H be as in Definition 11.3. Then there exists $\omega \in \mathbf{R}$ such that for all $r > 2n|\frac{1}{p} - \frac{1}{2}|$, $\{e^{isH}(\omega - H)^{-r}\}_{s \in \mathbf{R}}$ is an $(\omega - H)^{-r}$-regularized group on $L^p(\mathbf{R}^n)$ that is $O((1 + |s|^{2n|\frac{1}{p} - \frac{1}{2}|}))$.*

On $C_0(\mathbf{R}^n)$ or $BUC(\mathbf{R}^n)$, $\{e^{isH}(\omega - H)^{-r}\}_{s \in \mathbf{R}}$ is an $(\omega - H)^{-r}$-regularized group that is $O((1 + |s|^n))$, for all $r > n$.

Theorem 11.1 follows immediately from Theorem 10.1 and the following lemma.

Lemma 11.5. *Suppose* $1 \leq p \leq \infty$ *and* $n \in \mathbf{N}$. *Then*

$$\|e^{z\Delta}\|_p \leq \left(\frac{|z|}{Re(z)}\right)^{n|\frac{1}{p} - \frac{1}{2}|},$$

whenever $Re(z) > 0$.

Proof: The heat semigroup is a convolution operator with kernel $K_z(x) \equiv (4\pi z)^{-\frac{n}{2}} e^{-\frac{x^2}{4z}}$, that is,

$$e^{z\Delta} f = K_z * f,$$

for all $f \in L^p(\mathbf{R}^n)$, $Re(z) > 0$. Thus a direct computation shows that

$$\|e^{z\Delta}\|_1 = \|K_z\|_1 = \left(\frac{|z|}{Re(z)}\right)^{\frac{n}{2}}.$$

It is also well known that $\|e^{z\Delta}\|_2 = 1$, whenever $Re(z) > 0$.

Fix z in the open right half-plane. We now use the Riesz Convexity Theorem ([Du-S1, p. 523]), which asserts that $h(a) \equiv \log \|e^{z\Delta}\|_{\frac{1}{a}}$ is a convex function of a in $[0, 1]$. First consider $1 \leq p \leq 2$, and set $p = \frac{1}{a}$, so that a is in the interval $[\frac{1}{2}, 1]$. For such p, the convexity of h implies that

$$\log \|e^{z\Delta}\|_p \leq \alpha \log \|e^{z\Delta}\|_2 + \beta \log \|e^{z\Delta}\|_1 = \beta \log \|e^{z\Delta}\|_1,$$

where $\alpha + \beta = 1, \frac{1}{2}\alpha + \beta = a = \frac{1}{p}$. The last two equations imply that $\beta = \frac{2}{p} - 1$, thus

$$\|e^{z\Delta}\|_p \leq \|e^{z\Delta}\|_1^{\frac{2}{p} - 1} = \left(\frac{|z|}{Re(z)}\right)^{n(\frac{1}{p} - \frac{1}{2})},$$

proving the Lemma when $1 \leq p \leq 2$.

For $2 \leq p \leq \infty$, duality implies that, if $\frac{1}{p} + \frac{1}{q} = 1$, then

$$\|e^{z\Delta}\|_p = \|e^{z\Delta}\|_q \leq \left(\frac{|z|}{Re(z)}\right)^{n(\frac{1}{q}-\frac{1}{2})} = \left(\frac{|z|}{Re(z)}\right)^{n(\frac{1}{2}-\frac{1}{p})},$$

concluding the proof. ∎

For the proof of Theorem 11.4, we will need the following, from [Sim, Theorem B.7.1] and [Dav2, Theorem 9]; see also [P, Propositions 2.1 and 2.4].

Lemma 11.6. *Let H be as in Definition 11.3. Then there exist $\mu, c, a \in \mathbf{R}^+$ and a kernel $\tilde{K}(z, x, y)$ such that*

$$e^{z(H-\mu)} f(x) = \int_{\mathbf{R}^n} \tilde{K}(z, x, y) f(y) \, dy,$$

for $Re(z) > 0, f \in L^p(\mathbf{R}^n)(1 \leq p \leq \infty)$ and

$$|\tilde{K}(z, x, y)| \leq c(Re(z))^{-\frac{n}{2}} \exp\left(-Re\left(\frac{|x-y|^2}{az}\right)\right),$$

for $Re(z) > 0, x, y \in \mathbf{R}^n$.

Proof of Theorem 11.4: Let $\omega \equiv \max\{\|V_-\|_\infty, \mu\} + 1$.

We argue as in Theorem 11.1 with Δ replaced by $H + 1 - \omega$.

Since $V + \omega - 1$ is a real-valued nonnegative Kato perturbation, so that $-z(V + \omega - 1)$ is a dissipative Kato perturbation, $\|e^{z(H+1-\omega)}\|_2 \leq 1$, for all $Re(z) > 0$ (see [Go2, Corollary 6.8]).

By Lemma 11.6, $\|e^{z(H+1-\omega)}\|_1 \leq \|(Re(z))^{-\frac{n}{2}} \exp\left(-Re\left(\frac{|x|^2}{az}\right)\right)\|_1 \leq$ $c'\left(\frac{|z|}{Re(z)}\right)^n$, where c' is independent of z, so that we may argue exactly as in the proof of Theorem 11.1, to conclude that, for all $r > 2n|\frac{1}{p} - \frac{1}{2}|, \{e^{isH}(\omega - H)^{-r}\} = \{e^{is(\omega-1)} e^{is(H+1-\omega)}(1 - (H+1-\omega))^{-r}\}$ is a $O((1+ |s|^{2n|\frac{1}{p}-\frac{1}{2}|}))$ $(\omega - H)^{-r}$-regularized semigroup. ∎

XII. FUNCTIONAL CALCULUS FOR COMMUTING GENERATORS OF BOUNDED STRONGLY CONTINUOUS GROUPS

In this chapter, we give an example of how regularized semigroups may be used to construct functional calculi for unbounded operators, or families of commuting unbounded operators. The idea is to construct a regularized semigroup, $e^{tf(A)}g(A)$, for appropriate g, then define $f(A)$ as the generator. By "appropriate" g, intuitively, we want the map $w \mapsto e^{tf(w)}g(w)$ to be bounded on the spectrum of A; depending on the structure of A, we may impose stronger growth conditions on this map.

In this chapter, we will use the Fourier transform for our functional calculus construction. In Chapter XXII, we will use unbounded analogues of the Cauchy integral formula in a similar way.

Even for multivariable polynomials, p, choosing an appropriate domain for $p(B_1, \ldots, B_n)$, where $B_1, \ldots B_n$ are commuting operators, can be difficult. We shall see that regularized semigroups provide a very simple way of specifying a domain, namely, the domain of the generator of a regularized semigroup.

We will need some standard multivariable terminology. We will write $x = (x_1, \ldots x_n)$, for vectors in $\mathbf{R}^n$, $\alpha = (\alpha_1, \ldots \alpha_n)$ for vectors in $(\mathbf{N} \cup \{0\})^n$. We will write $x^\alpha \equiv x_1^{\alpha_1} \cdots x_n^{\alpha_n}, |x|^2 \equiv \sum_{k=1}^n |x_k|^2, |\alpha| \equiv \sum_{k=1}^n \alpha_k$.

Terminology 12.1. Throughout this chapter, and the next two, $iB_1, \ldots, iB_n$ will be commuting generators of bounded strongly continuous groups of operators. We will write

$$B \equiv (B_1, \ldots, B_n), \quad B^\alpha \equiv B^{\alpha_1} \cdots B^{\alpha_n}.$$

Let $\mathcal{F}$ be the Fourier transform. We define $\mathcal{A}$ to be the set of all inverse Fourier transforms of L^1 functions, that is, $\mathcal{A} \equiv \{f \in C_0(\mathbf{R}^n) \mid \mathcal{F}f \in L^1(\mathbf{R}^n)\}$.

We will write $e^{i(x \cdot B)}$ for $\prod_{k=1}^n e^{ix_k B_k}$, where $\{e^{itB_k}\}_{t \in \mathbf{R}}$ is the strongly continuous group generated by iB_k. We will use the following well-known functional calculus. Define a bounded operator $f(B)$ by

$$f(B) \equiv (2\pi)^{-\frac{n}{2}} \int_{\mathbf{R}^n} e^{i(x \cdot B)} \mathcal{F}f(x) \, dx, \qquad (12.2)$$

whenever $f \in \mathcal{A}$. Note that this is essentially the Fourier inversion theorem.

The map $f \mapsto f(B)$ has the following properties.

(1) It is an algebra homomorphism.

(2) There exists $M < \infty$ such that

$$\|f(B)\| \leq M\|\mathcal{F}f\|_1,$$

for all $f \in \mathcal{A}$.

(3)

$$\prod_{j=1}^{n}(\lambda_j - B_j)^{-\alpha_j} = \left(\prod_{j=1}^{n}(\lambda_j - x_j)^{-\alpha_j}\right)(B),$$

for any $\alpha \in \mathbf{N}^n, Im(\lambda_j) \neq 0$.

Let $-|B|^2$ be the generator of the strongly continuous semigroup

$$e^{-t|B|^2} \equiv f_t(B), \ f_t(x) \equiv e^{-t|x|^2}.$$

Then in the usual way for generators of strongly continuous semigroups (see any of the references for strongly continuous semigroups), we have fractional powers $(1 + |B|^2)^{-r}$, for $r > 0$; in this case, they are given by

$$(1 + |B|^2)^{-r} = ((1 + |x|^2)^{-r})(B),$$

defined by (12.2).

For $k \in \mathbf{N}$, we will denote by $BC^k(\mathbf{R}^n)$ the algebra of complex-valued functions on $\mathbf{R}^n$, with bounded continuous derivatives, up to order k, with norm

$$\|f\|_{BC^k(\mathbf{R}^n)} \equiv \sum_{|\alpha| \leq k} \|D^\alpha f\|_\infty.$$

Proposition 12.3. Let $k \equiv 1 + [\frac{n}{2}]$. For any $s > \frac{n}{4}$, there exists $M(s) < \infty$ such that, for any $f \in BC^k(\mathbf{R}^n)$, $(f(x)(1 + |x|^2)^{-s}) \in \mathcal{A}$, and

$$\| (f(x)(1 + |x|^2)^{-s})(B)\| \leq M(s)\|f\|_{BC^k(\mathbf{R}^n)}.$$

Remark 12.4. The map $\Lambda f \equiv (f(x)(1 + |x|^2)^{-s})(B)$ defines a multi-variable example of a $((1 + |B|^2)^{-s})$ *regularized functional calculus* (see Definition 22.11).

Definition 12.5. For $f \in C^\infty(\mathbf{R}^n)$, let $\mathcal{A}(f)$ be the set of all real, positive bounded functions $g \in C^\infty(\mathbf{R}^n)$, such that $z \mapsto e^{zf}g$ defines an entire map from the complex plane into $BC^k(\mathbf{R}^n)$).

Before proceeding further, we should verify that this set is nonempty.

Lemma 12.6. *For any* $f \in C^\infty(\mathbf{R}^n)$, $\mathcal{A}(f)$ *is nonempty.*

Definition 12.7. *For any* $f \in C^\infty(\mathbf{R}^n)$, $g \in \mathcal{A}(f)$, *complex* z, *let*

$$\mathcal{G}_{f,g,z}(x) \equiv e^{zf(x)}g(x)(1+|x|^2)^{-s},$$

where $s \equiv 1 + [\frac{n}{4}]$.
For any $z \in \mathbf{C}$, define

$$W_{f,g}(z) \equiv \mathcal{G}_{f,g,z}(B).$$

Theorem 12.8. *The family* $\{W_{f,g}(z)\}_{z\in\mathbf{C}}$ *is an entire* $W_{f,g}(0)$-*regularized group.*

Propositon 12.9. *If* $f \in \mathcal{A} \cap C^\infty(\mathbf{R}^n)$, *then* $f(B)$ *is the generator of* $\{W_{f,g}(z)\}_{z\in\mathbf{C}}$, *for any* $g \in \mathcal{A}(f)$.

Definition 12.10. *For* $f \in \mathbf{C}^\infty(\mathbf{R}^n)$, $f(B)$ *is defined to be the generator of* $\{W_{f,g}(z)\}_{z\in\mathbf{C}}$, *for any* $g \in \mathcal{A}(f)$.
The operator is well-defined, that is, the definition is independent of which $g \in \mathcal{A}(f)$ is chosen, for the following reason. Temporarily write $f(j, B)$ $(j = 1, 2)$ for the generator of $\{W_{f,g_j}(z)\}_{z\in\mathbf{C}}$. Then by Proposition 3.10, both $f(1, B)$ and $f(2, B)$ generate the regularized group $\{W_{f,g_1g_2}(z)\}_{z\in\mathbf{C}}$, thus are equal.
Proposition 12.9 shows that this definition is consistent with (12.2).

In the following theorem, note that any N^{th} degree polynomial will satisfy (1).

Theorem 12.11. *Suppose* $f \in C^\infty(\mathbf{R}^n)$, $k \equiv 1 + [\frac{n}{2}], w \in \mathbf{R}, N \in \mathbf{N}, M \geq 0$ *such that for all* $x \in \mathbf{R}^n$,

(1)
$$|(D^\alpha f)(x)| \leq M(1+|x|)^N,$$

whenever $|\alpha| \leq k$; and

(2)
$$Re(f(x)) \leq \omega.$$

Then, for all $r > (\frac{k}{2}(N-1) + \frac{n}{4})$, $f(B)$ generates a norm continuous $(1 + |B|^2)^{-r}$-regularized semigroup that is $O(e^{\omega t})$.

The following states two ways in which our map $f \mapsto f(B)$ is an algebra homomorphism. The consistency of Definitions (12.2) and 12.10

has already been established. When p is a polynomial of degree N in n variables, $p(x) \equiv \sum_{|\alpha| \leq N} c_\alpha x^\alpha$, then we have the natural definition

$$p(B) \equiv \sum_{|\alpha| \leq N} c_\alpha B^\alpha.$$

We must verify that this is consistent with Definition 12.10, at least on a dense set.

Theorem 12.12.

(1) *If h and fh are in A and f and h are in $C^\infty(\mathbf{R}^n)$, then $Im(h(B)) \subseteq \mathcal{D}(f(B))$, with*

$$(fh)(B) = f(B)h(B).$$

(2) *Suppose $p(x) \equiv \sum_{|\alpha| \leq N} c_\alpha x^\alpha$ and $2\ell > N + \frac{n}{2}$. Then $\mathcal{D}(|B|^{2\ell}) \subseteq \mathcal{D}(p(B))$ and*

$$p(B)x \equiv \sum_{|\alpha| \leq N} c_\alpha B^\alpha x,$$

for all $x \in \mathcal{D}(|B|^{2\ell})$.

Lemma 12.13.

$$I(s) \equiv \int_{\mathbf{R}^n} (1 + |x|^2)^{-s} \, dx < \infty,$$

whenever $s > \frac{n}{2}$.

Lemma 12.14. If $k \equiv 1 + [\frac{n}{2}]$, then there exists $M_1 < \infty$ such that

$$\int_{\mathbf{R}^n} |(\mathcal{F}f)(x)| \, dx \leq M_1 \sum_{|\alpha| \leq k} \|D^\alpha f\|_2,$$

for all $f \in H^k(\mathbf{R}^n)$.

Proof of Proposition 12.3: By Lemmas 12.13 and 12.14, and (2) of (12.2), there exists a constant $M_2 < \infty$ such that

$$\|(f(x)(1 + |x|^2)^{-s})(B)\| \leq M_1 \|\mathcal{F}\left(f(x)(1 + |x|^2)^{-s}\right)\|_1$$

$$\leq M_1 \sum_{|\alpha| \leq k} \|D^\alpha \left(f(x)(1 + |x|^2)^{-s}\right)\|_2$$

$$\leq M_2 \|(1 + |x|^2)^{-s}\|_2 \sum_{|\alpha| \leq k} \|D^\alpha f\|_\infty$$

$$= M_2 \left(I(2s)\right)^{\frac{1}{2}} \|f\|_{BC^k(\mathbf{R}^n)}.$$

82

Proof of Lemma 12.6: For any $F \in C^\infty(\mathbf{R}^n)$, $1 \le i \le n$, define

$$(J_i F)(x) \equiv \int_{x_i}^\infty F(x_1, \ldots x_{i-1}, s, x_{i+1}, \ldots x_n) \, ds \ (x_i \ge 0),$$

$$(J_i F)(x) \equiv \int_{-\infty}^{x_i} F(x_1, \ldots x_{i-1}, s, x_{i+1}, \ldots x_n) \, ds \ (x_i < 0).$$

Fix $f \in C^\infty(\mathbf{R}^n)$. Let $k \equiv 1 + [\frac{n}{2}]$, as in Proposition 12.3. For $|\alpha| \le k$, define

$$F_\alpha(s) \equiv e^{-|s|^2} \sup_{|y| \le |s|} |D^\alpha f(y)|^2.$$

Define

$$g \equiv \left(\prod_{i=1}^n J_i^k \right) \left(\prod_{|\alpha| \le k} e^{-F_\alpha} \right).$$

Then a calculation shows that $g \in \mathcal{A}(f)$. ∎

Lemma 12.15. For any $f \in C^\infty(\mathbf{R}^n)$, $g \in \mathcal{A}(f)$, the map $z \mapsto W_{f,g}(z)$ is entire, with

$$(\frac{d}{dz})^k W_{f,g}(z) = (f^k \mathcal{G}_{f,g,z})(B),$$

for all complex $z, k \in \mathbf{N}$.

Proof: This follows from (1) of (12.2), and Proposition 12.3. ∎

Lemma 12.16. Suppose $f \in C^\infty(\mathbf{R}^n)$ and $g \in \mathcal{A}(f)$. Then $W_{f,g}(0)$ is injective.

Proof: Define, for any complex z,

$$T(z) \equiv (e^{z \ln g(x)} (1 + |x|^2)^{-s})(B),$$

where s is as in Definition 12.7.

By Proposition 12.3, this defines an entire family of bounded operators. Since $T(0) = (1 + |B|^2)^{-\ell}$, which is injective, the same proof as that given for Lemma 8.8 implies that $T(z)$ is injective, for all complex z. In particular, $W_{f,g}(0) = T(1)$ is injective. ∎

Proof of Theorem 12.8: By Lemma 12.16, $C \equiv W_{f,g}(0)$ is injective. By Lemma 12.15, the map $z \mapsto W_{f,g}(z)$ is entire. By (12.2)(1), $\{W_{f,g}(z)\}_{z \in \mathbf{C}}$ is an entire C-regularized group. ∎

Proof of Proposition 12.9: By Lemma 12.15,

$$(\frac{d}{dz}) W_{f,g}(z)|_{z=0} = (\mathcal{G}_{f,g,0} f)(B) = W_{f,g}(0) f(B),$$

83

by (12.2)(1). The result now follows from the definition of the generator of a regularized semigroup. ∎

The following lemma is a straightforward calculation, using the product rule.

Lemma 12.17. For $k \in \mathbf{N}, s > 0, f$ as in (1) of Theorem 12.11 fixed, there exists $K(s) < \infty$ such that

$$\left\| \left(e^{tf(x)}(1 + |x|^2)^{-s} \right) \right\|_{BC^k(\mathbf{R}^n)} \leq K(s) \|(1 + |x|)^{k(N-1)-2s} e^{tf(x)} \|_{\infty}.$$

Proof of Theorem 12.11: By translating, we may assume $\omega = 0$. Fix $r > \frac{k}{2}(N-1) + \frac{n}{4}$. Let $\mathcal{G}_t(x) \equiv e^{tf(x)}(1 + |x|^2)^{-r}$, for $t \geq 0, x \in \mathbf{R}^n$.
We claim that $\{\mathcal{G}_t(B)\}_{t \geq 0}$ is the desired regularized semigroup.
Choose $s_1 > \frac{n}{4}, s_2 > \frac{k}{2}(N-1)$ so that $s_1 + s_2 = r$.
By Proposition 12.3 and Lemma 12.17,

$$\begin{aligned}
\|\mathcal{G}_t(B)\| &= \|(e^{tf(x)}(1 + |x|^2)^{-s_2})(B)(1 + |B|^2)^{-s_1}\| \\
&\leq M(s_1)\|(e^{tf(x)}(1 + |x|^2)^{-s_2})\|_{BC^k(\mathbf{R}^n)} \\
&\leq M(s_1)K(s_2)\|(1 + |x|)^{(k(N-1)-2s_2)} e^{tf(x)}\|_{\infty} \\
&\leq M(s_1)K(s_2).
\end{aligned}$$

Thus $\{\|\mathcal{G}_t(B)\|\}_{t \geq 0}$ is uniformly bounded. The same argument applied to $\|\mathcal{G}_t(B) - \mathcal{G}_s(B)\|$ implies that $\{\mathcal{G}_t(B)\}_{t \geq 0}$ is norm continuous.

Since $f \mapsto f(B)$ is an algebra homomorphism, $\{\mathcal{G}_t(B)\}_{t \geq 0}$ is a $\mathcal{G}_0(B)$-regularized semigroup. By (3) of (12.2), $\mathcal{G}_0(B) = (1 + |B|^2)^{-r}$.

Thus $\{\mathcal{G}_t(B)\}_{t \geq 0}$ is a $(1 + |B|^2)^{-r}$-regularized semigroup. Choose $g \in \mathcal{A}(f)$. By (12.2)(1), $g(B)(1 + |B|^2)^{-s}\mathcal{G}_t(B) = (1 + |B|^2)^{-r}W_{f,g}(t)$ (see Definitions 12.7 and 12.10). By Proposition 3.10, $f(B)$ is the generator of $\{\mathcal{G}_t(B)\}_{t \geq 0}$. ∎

Proof of Theorem 12.12: For (1), we use (1) of (12.2) and Proposition 12.3, as follows. Suppose $g \in \mathcal{A}(f)$. Then

$$\frac{d}{dz}W_{f,g}(z)|_{z=0}h(B) = (\mathcal{G}_{f,g,0}f)(B)h(B) = (\mathcal{G}_{f,g,0}fh)(B)$$

$$= W_{f,g}(0)(fh)(B).$$

The definition of $f(B)$ as the generator of a $W_{f,g}(0)$- regularized group now concludes the proof of (1).

For (2), we apply (1) with $h(x) \equiv (1 + |x|^2)^{-\ell}$. By Proposition 12.3, h and ph are in $\mathcal{A}$. Thus, by (1), $\mathcal{D}(|B|^{2\ell}) = Im((1 + |B|^2)^{-\ell}) \subseteq \mathcal{D}(p(B))$, with

$$p(B)(1 + |B|^2)^{-\ell} = \left(p(x)(1 + |x|^2)^{-\ell}\right)(B) = \sum_{|\alpha| \leq N} c_\alpha B^\alpha (1 + |B|^2)^{-\ell},$$

by (3) of (12.2). ∎

XIII. PETROVSKY CORRECT MATRICES OF GENERATORS OF BOUNDED STRONGLY CONTINUOUS GROUPS

In considering the system of differential equations in a Banach space,

$$\frac{d}{dt}\vec{u}(t) = (p_{i,j}(B_1,\dots,B_n))\vec{u}(t)\,(t \ge 0),\ \vec{u}(0) = \vec{x}, \qquad (13.1)$$

where $(p_{i,j})$ is a matrix of polynomials in n variables and $iB_1,\dots,iB_n$ is a family of commuting generators of bounded strongly continuous groups of operators, intuitively, one wants to take the exponential of the matrix of operators $(p_{i,j}(B))$. Since the spectrum of B_k is real, for all k, this means we want the complex matrix $(p_{i,j}(x))$ to be contained in a left half-plane, "in some sense," for all x in $\mathbf{R}^n$. The strongest sense is to have the *numerical range* of the complex matrix contained in a half-plane $Re(z) \le \omega$, since this implies that the matrix-valued exponential $e^{t(p_{i,j}(x))}$ has norm less than or equal to $e^{t\omega}$, for all $t \ge 0, x \in \mathbf{R}^n$. A sense that is easier to calculate, or at least estimate, is to have the *spectrum* of $(p_{i,j}(x))$ contained in a left half-plane, for all $x \in \mathbf{R}^n$. The system (13.1) is then said to be *Petrovsky correct* (see [G-S]).

The intuition we just described is not quite true, in general, without some modification. When $B_k \equiv \frac{i\partial}{\partial x_k}$, on $L^q(\mathbf{R}^n)(1 \le q < \infty)$, then $(p_{i,j}(B_1,\dots,B_n))$ generates a strongly continuous semigroup, whenever the numerical range of $(p_{i,j}(x))$ is contained in the left half-plane, if and only if $q = 2$. This is true even for a one-by-one matrix: take $p(x) \equiv i|x|^2$, so that $p(B)$ is the Schrödinger operator $i\triangle$; then it is well known that $p(B)$ generates a strongly continuous semigroup if and only if $q = 2$.

Higher order abstract Cauchy problems are a special case of (13.1), via the usual matrix reduction to a first order problem. The obvious choice for B_k is $\frac{i\partial}{\partial x_k}$, on some space of functions on $\mathbf{R}^n$, so that (13.1) becomes an arbitrary system of constant coefficient differential equations. But certainly other choices are possible, such as $\frac{i\partial}{\partial x_k}$ on a space of functions on some bounded domain, with appropriate boundary conditions.

In this chapter, we use regularized semigroups to describe simply and explicitly how a perturbation of the intuition described above is true, when the matrix $(p_{i,j}(x))$ is Petrovsky correct. The operator $(p_{i,j}(B_1,\dots,B_n))$ then generates a $(1 + |B|^2)^{-r}$- regularized semigroup, where

$$|B|^2 \equiv \sum_{k=1}^n |B_k|^2$$

and r is a constant depending on n, the order of the matrix and the maximum value of the degree of $p_{i,j}$ (Theorem 13.9). When $(p_{i,j}(x))$ has numerical range contained in a left half-plane, independently of x, then the dependence on the order of the matrix disappears (Theorem 13.4).

Our goal in this chapter has been to obtain results involving nothing more than the most simple-minded spectral intuition, that is, the behaviour of the complex-valued matrices $(p_{i,j}(x))$, for x in $\mathbf{R}^n$.

We will use the same terminology as in the previous chapter. Throughout this chapter, m and n are fixed natural numbers.

Matrix Operator Terminology 13.2 We give X^m the Banach space norm

$$\|(x_1, \ldots x_m)\| \equiv \sum_{k=1}^{m} \|x_k\|.$$

Throughout this chapter and the next, $\mathcal{M} \equiv (p_{i,j})_{i,j=1}^{m}$ will be an $m \times m$ matrix of polynomials. The number N will be $\max_{i,j}\{$ degree of $p_{i,j}\}$. We define the operator $\mathcal{M}(B)$, on X^m, by

$$\mathcal{M}(B) \equiv (p_{i,j}(B)), \ \mathcal{D}(\mathcal{M}(B)) \equiv \mathcal{D}(|B|^{2\ell})^m, \ell \equiv 1 + [\frac{1}{2}(N + \frac{n}{2})].$$

(See Theorem 12.2(2).)

If $\mathcal{F}f_{i,j} \in L^1(\mathbf{R}^n)$, and $\mathcal{G} = (f_{i,j})$, we define the operator $\mathcal{G}(B) \equiv (f_{i,j}(B)) \in B(X^m)$ similarly.

Note that

$$\|\mathcal{G}(B)\| \leq \sum_{1 \leq i,j \leq m} \|f_{i,j}(B)\|.$$

When A is an operator on X, we will also write A for AI_m, the operator on X^m with domain $(\mathcal{D}(A))^m$.

Terminology 13.3. If M is an $m \times m$ matrix of complex numbers, we will write $n.r.(M)$ for the *numerical range of* M,

$$n.r.(M) \equiv \{< x, Mx > \mid x \in \mathbf{R}^m, |x| = 1\},$$

where $< \cdot >$ is the inner product in $\mathbf{C}^m$, $< x, y > \equiv \sum_{i=1}^{m} x_i \overline{y_i}$.

Theorem 13.4. *Suppose* $w \in \mathbf{R}, k \equiv ([\frac{n}{2}] + 1)$ *and for all* $x \in \mathbf{R}^n$,

$$n.r.(\mathcal{M}(x)) \subseteq \{z \in \mathbf{C} \mid Re(z) \leq w\}.$$

Then, for all $r > (\frac{k}{2}(N-1) + \frac{n}{4})$, an extension of $\mathcal{M}(B)$ generates a norm continuous $O(e^{wt})\, (1+|B|^2)^{-r}$-regularized semigroup, that leaves $\mathcal{D}(\mathcal{M}(B))$ invariant.

Remark 13.5. It is interesting that this result is independent of m, the size of the matrix (compare with Theorem 12.11).

Example 13.6. In Theorem 13.4, let $B_k \equiv iD_k, D \equiv i(D_1, \ldots, D_n)$, $X \equiv$ the Sobolev space $W^{\ell,q}(\mathbf{R}^n)(1 \le q < \infty, \ell+1 \in \mathbf{N})$. Then since $W^{j,q}(\mathbf{R}^n) \subseteq \mathcal{D}(\Delta^{\frac{i}{2}}) = Im((1+\Delta)^{-\frac{i}{2}})$, Theorems 13.4, 3.8(3) and 5.16 give us the following result about systems of constant coefficient Cauchy problems,

$$\frac{d}{dt}\vec{u}(t,\vec{f})(x) = (p_{i,j}(D))\vec{u}(t,\vec{f})(x) \quad \vec{u}(0,\vec{f})(x) = \vec{f}(x) \quad (t \ge 0, x \in \mathbf{R}^n).$$
(13.7)

Corollary. *Suppose $w \in \mathbf{R}$ and for all $x \in \mathbf{R}^n$,*

$$n.r.(p_{i,j}(x)) \subseteq \{z \in \mathbf{C} \,|\, Re(z) \le w\}$$

and $j \equiv \frac{n}{2} + N + (N-1)[\frac{n}{2}]$. Then for all nonnegative integers $\ell, 1 \le q < \infty$, there exists $Z_{\ell,q}$, a Banach space, such that

$$(W^{j+\ell,q}(\mathbf{R}^n))^m \hookrightarrow Z_{\ell,q} \hookrightarrow (W^{\ell,q}(\mathbf{R}^n))^m,$$

and (13.7) is well-posed on $Z_{\ell,q}$, that is, for all $\vec{f} \in Z_{\ell,q}$, there exists a unique mild solution of (13.7) in $Z_{\ell,q}$ and $\vec{u}(t, f_n) \to 0$ in $Z_{\ell,q}$, uniformly in t on compact subsets, whenever $f_n \to 0$ in $Z_{\ell,q}$.

Of course, similar results hold for $X = C_0(\mathbf{R}^n), C^\alpha(\mathbf{R}^n)$ or $C_0^k(\mathbf{R}^n)$, since translation is uniformly bounded and strongly continuous on these spaces.

Definition 13.8. The matrix of polynomials $\mathcal{M}$ is said to be *Petrovsky correct* (see [G-S]) if there exists $w \in \mathbf{R}$ such that for all $x \in \mathbf{R}^n$,

$$\sigma(\mathcal{M}(x)) \subseteq \{z \in \mathbf{C} \,|\, Re(z) \le w\}.$$

Theorem 13.9. *Suppose $\mathcal{M}$ is Petrovsky-correct, $k \equiv ([\frac{n}{2}]+1)$ and $r > (\frac{k}{2}(N-1) + \frac{n}{4} + \frac{N}{2}(m-1))$. Then an extension of $\mathcal{M}(B)$ generates an exponentially bounded norm continuous $(i+|B|^2)^{-r}$-regularized semigroup, that leaves $\mathcal{D}(\mathcal{M}(B))$ invariant.*

Specifically, if w is as in Definition 13.8, then this regularized semigroup is $O(e^{(w+\epsilon)t})$, for all $\epsilon > 0$

Example 13.11. As in Example 13.6, the obvious application is to Petrovsky-correct systems of constant coefficient Cauchy problems.

88

Corollary. *Suppose $(p_{i,j})$ is Petrovsky-correct and $j \equiv \frac{n}{2} + Nm + (N-1)[\frac{n}{2}]$. Then for all nonnegative integers $\ell, 1 \leq q < \infty$, there exists $Z_{\ell,q}$, a Banach space, such that*

$$(W^{j+\ell,q}(\mathbf{R}^n))^m \hookrightarrow Z_{\ell,q} \hookrightarrow (W^{\ell,q}(\mathbf{R}^n))^m,$$

and (13.7) is well-posed on $Z_{\ell,q}$, that is, for all $\vec{f} \in Z_{\ell,q}$, there exists a unique mild solution of (13.7) in $Z_{\ell,q}$ and $\vec{u}(t, f_n) \to 0$ in $Z_{\ell,q}$, uniformly in t on compact subsets, whenever $f_n \to 0$ in $Z_{\ell,q}$.

Of course, similar results hold for $X = C_0(\mathbf{R}^n), C^\alpha(\mathbf{R}^n)$ or $C_0^k(\mathbf{R}^n)$, since translation is uniformly bounded and strongly continuous on these spaces.

Example 13.12. The equation describing sound propagation in a viscous gas is (see [G-S, Example 3, page 134])

$$\frac{\partial^2 u}{\partial t^2} = 2\frac{\partial^3 u}{\partial t \partial x^2} + \frac{\partial^2 u}{\partial x^2}.$$

After the usual matrix reduction, this becomes

$$\frac{d}{dt}\vec{u} = \mathcal{M}(D)\vec{u},$$

where $D \equiv i\frac{d}{dx}, \mathcal{M}(x) \equiv \begin{bmatrix} 0 & 1 \\ (-x^2) & (-2x^2) \end{bmatrix}$. Since the spectrum of $\mathcal{M}(x)$ is $-x^2(1 \pm \sqrt{x^2 - 1})$, which has real part bounded above, for x real, this is a Petrovsky-correct system, so that Theorem 13.9 implies that an extension of $\mathcal{M}(D)$ generates an exponentially bounded $(1 + \Delta)^{-r}$-regularized semigroup, for all $r > \frac{7}{4}$.

The following lemma is a straightforward calculation, using the product rule and the fact that $|x^\alpha| \leq |x|^\alpha$.

For matrix-valued functions $\mathcal{G} \equiv (f_{i,j})$, with $f_{i,j} \in BC^k(\mathbf{R}^n)$, we define

$$\|\mathcal{G}\|_{BC^k(\mathbf{R}^n, B(\mathbf{C}^m))} \equiv \sum_{1 \leq i,j \leq m} \|f_{i,j}\|_{BC^k(\mathbf{R}^n)}.$$

Lemma 13.13. *For $k \in \mathbf{N}, r > 0, p_{i,j}$ fixed, there exists $M < \infty$ such that*

$$\left\| \left(e^{t(p_{i,j}(x))_{i,j=1}^m}(1 + |x|^2)^{-r} \right) \right\|_{BC^k(\mathbf{R}^n, B(\mathbf{C}^m))}$$

$$\leq M \sup_{x \in \mathbf{R}^n} \left[(1 + |x|)^{k(N-1) - 2r} \|e^{t(p_{i,j}(x))_{i,j=1}^m}\|_{B(\mathbf{C}^m)} \right].$$

89

Proof of Theorem 13.4: Let $\mathcal{G}_t(x) \equiv e^{t(p_{i,j}(x))}(1 + |x|^2)^{-r}$, for $t \geq 0$, $x \in \mathbf{R}^n$. As in the proof of Theorem 12.11, with Lemma 13.13 replacing Lemma 12.17, $\{\mathcal{G}_t(B)\}_{t \geq 0}$ is a $(1 + |B|^2)^{-r}$-regularized semigroup.

Using dominated convergence and Proposition 12.3, a calculation shows that $\mathcal{G}_t(B)(1 + |B|^2)^{-\ell}$ is (norm) differentiable at $t = 0$, and

$$\frac{d}{dt}\mathcal{G}_t(B)(1 + |B|^2)^{-\ell}|_{t=0} = (1 + |B|^2)^{-r}\mathcal{M}(B)(1 + |B|^2)^{-\ell}.$$

Since $\mathcal{D}(\mathcal{M}(B)) = Im((1 + |B|^2)^{-\ell})$, this implies that an extension of $\mathcal{M}(B)$ generates the regularized semigroup. The invariance of $\mathcal{D}(\mathcal{M}(B))$ under the regularized semigroup is clear from the fact that $\mathcal{G}_t(B)$ and $(1 + |B|^2)^{-\ell}$ commute. ∎

Lemma 13.13 tells us that the proof of Theorem 13.9 should be the same as the proof of Theorem 13.4, if we can estimate $\|e^{t(p_{i,j}(x))^m_{i,j=1}}\|$. Thus we need some lemmas about matrices. Although the following may be well known, we include their proofs for completeness.

Lemma 13.14. *For any $m \in \mathbf{N}$, there exists $K_1(m)$ such that*

$$\det(M)\|M^{-1}\| \leq K_1(m)\|M\|^{m-1},$$

for all $m \times m$ invertible matrices M.

Proof: Look at the cofactor expansion for M^{-1}. ∎

Lemma 13.15. *For any $m \in \mathbf{N}$, there exists $K_2(m)$ with the following property. Whenever $\epsilon > 0$ and M is an $m \times m$ matrix with*

$$\sigma(M) \subseteq \{z \in \mathbf{C} \,|\, Re(z) < -\epsilon\},$$

then for all $t \geq 0$,

$$\|e^{tM}\| \leq \frac{K_2(m)}{\epsilon^{m-1}}\|M\|^{m-1}.$$

Proof: Let $\{b_k\}^m_{k=1}$ be the (not necessarily distinct) eigenvalues of M.

It is elementary to verify that there exists Γ, a contour of length less than $3\pi m^2 \epsilon$, contained in the left half-plane $Re(z) < 0$, that surrounds

$\sigma(M)$, such that $|z - b_k| \geq \frac{\epsilon}{2}$, for all $z \in \Gamma$. Then, using Lemma 13.14,

$$2\pi \|e^{tM}\| = \left\| \int_\Gamma e^{tz}(z - M)^{-1} dz \right\|$$

$$\leq \int_\Gamma \|(z - M)^{-1}\| d|z|$$

$$\leq K_1(m) \int_\Gamma \frac{\|(z - M)\|^{m-1}}{\prod_{k=1}^m |(z - b_k)|} d|z|$$

$$\leq \left(\frac{2}{\epsilon}\right)^m K_1(m) \int_\Gamma (|z| + \|M\|)^{m-1} d|z|$$

$$\leq \frac{2^m 3\pi m^2}{\epsilon^{m-1}} K_1(m)((\sup_{z \in \Gamma} |z|) + \|M\|)^{m-1}.$$

∎

Proof of Theorem 13.9: This is the same as the proof of Theorem 13.4, using Lemma 13.15 to replace the numerical range condition of Theorem 13.4. ∎

91

XIV. ARBITRARY MATRICES OF GENERATORS OF BOUNDED STRONGLY CONTINUOUS GROUPS

In this chapter, we obtain a result that, like the results in Chapters VIII and IX, may not seem intuitive. For *any* matrix $(p_{i,j}(x))$, there exists C, with dense image, such that $(p_{i,j}(B_1, \ldots, B_n))$ generates a C-regularized semigroup (Theorem 14.1). In particular, for any system of constant coefficient Cauchy problems or any constant coefficient higher order Cauchy problem, this produces entire solutions for all initial data in a dense set, along with the information about well-posedness in Chapter IV (Theorem 4.13).

Of course, the choice of C is more extreme than appears in Chapter XIII, since the matrix exponential $e^{t(p_{i,j}(x))}$ is now growing like e^{tx^N}, so that the formal exponential $e^{t(p_{i,j})}(B)$ now requires more smoothing to become a bounded operator.

There is a simple spectral intuition. Since $e^{t(p_{i,j}(x))}e^{-|x|^{2N}}$ is exponentially decaying, as $|x|$ goes to infinity, we choose $C \equiv e^{-|B|^{2N}}$.

We use the same terminology as in the previous chapter.

Theorem 14.1. *Let $(p_{i,j})$ be a matrix of polynomials as in 13.2. Then an extension of $\mathcal{M}(B)$ generates an entire $e^{-|B|^{2N}}$-regularized group that leaves $\mathcal{D}(\mathcal{M}(B))$ invariant.*

Remark 14.2. Note that, by Lemma 8.8, $e^{-|B|^{2N}}$ is injective and has dense range.

Example 14.3. A special case of (13.1), after the usual matrix reduction, is the higher-order abstract Cauchy problem

$$u^{(n)}(t) = A(u(t)) (t \geq 0), \ u^{(k)}(0) = x_k (0 \leq k < n).$$

It is well known (see [Hil]) that, when $n > 2$, this is well-posed, in the sense of strongly continuous semigroups, only when A is bounded. Thus Theorem 14.1, which asserts that this is well-posed in the sense of regularized semigroups, so that it is well-posed in the sense of strongly continuous semigroups on a Frechet space continuously embedded between a dense Banach subspace and X, may seem surprising.

Example 14.4. With $B_k \equiv i\frac{\partial}{\partial x_k}$, on $L^p(\mathbf{R}^n)$, Theorem 14.1 is saying that every system of constant coefficient Cauchy problems has a unique solution, for all initial data in a dense subset of $L^p(\mathbf{R}^n)$, with the well-posedness of Chapter IV.

Proof of Theorem 14.1: By Corollary 7.10, it is sufficient to show that

$$\sum_k \frac{s^k}{k!} \|(p_{i,j}(B))^k e^{-|B|^{2N}}\| < \infty,$$

for all positive s. But this follows from Proposition 12.3 and the fact that $\sum_k \frac{s^k}{k!}(\sqrt{k})^k$ converges, for any positive s. ∎

XV. MORE EXAMPLES OF REGULARIZED SEMIGROUPS

Probably the most well-known example of an unbounded semigroup of operators is $\{B^t\}_{t\geq 0}$, the family of fractional powers of a fixed operator B. The standard hypothesis on B is that of Definition 8.1. In Example 15.1, we show that there exists bounded, injective C, with dense range, such that $\{B^tC\}_{t\geq 0}$ is a C-regularized semigroup. We also consider sums of commuting generators of integrated semigroups (Example 15.2; see Definition 18.1) and the linearized form of the Ricatti operator (Example 15.3).

Example 15.1. Suppose B is closed, $\mathcal{D}(B)$ is dense and $(-\infty, 0) \subseteq \rho(B)$, with $\{\|r(r+B)^{-1}\| \,|\, r > 0\}$ bounded. For $t > 0$, let B^t be the usual fractional power of B (see Definition 8.1). Then, by the properties of strongly continuous holomorphic semigroups, $\{B^t e^{-B^{\frac{1}{2}}}\}_{t\geq 0}$ is an $e^{-B^{\frac{1}{2}}}$-regularized semigroup (note that $-B^{\frac{1}{2}}$ is constructed in such a way that it generates a strongly continuous holomorphic semigroup $\{e^{-sB^{\frac{1}{2}}}\}_{s\geq 0}$).

We remark that this regularized semigroup is far from being exponentially bounded, in general. If we choose

$$(Bf)(x) \equiv x^2 f(x),$$

on $C_0(\mathbf{R})$, with maximal domain, then a calculus calculation shows that for any $t > 0$,

$$\|B^t e^{-B^{\frac{1}{2}}}\| = \left(\frac{2t}{e}\right)^{2t}.$$

Example 15.2. See Definition 18.1 for the definition of an integrated semigroup. When B_1 and B_2 generate commuting exponentially bounded integrated semigroups, even if $\rho(B_1 + B_2)$ is nonempty, it is difficult to write down an integrated semigroup generated by $B_1 + B_2$.

The simplest way to deal with sums of generators of integrated semigroups is with regularized semigroups.

Theorem. *Suppose B_1 and B_2 generate exponentially bounded n-times integrated semigroups that commute and $Ax \equiv B_1x + B_2x$, $\mathcal{D}(A) \equiv \mathcal{D}(B_1B_2)$. Then an extension of A generates an $(r - B_1)^{-n}(r - B_2)^{-n}$-regularized semigroup that leaves $\mathcal{D}(A)$ invariant.*

Proof: By Theorem 18.3, there exists r, in $\rho(B_1) \cap \rho(B_2)$, such that, for $i = 1, 2$, B_i generates an exponentially bounded $(B_i - r)^{-n}$-regularized semigroup, $W_i(t)$. Let $W(t) \equiv W_1(t)W_2(t)$. This is a $(B_1 - r)^{-n}(B_2 - $

$r)^{-n}$-regularized semigroup. The following argument shows that an extension of A generates $W(t)$.

Suppose x is in $\mathcal{D}(A)$. For $t > 0$,

$$\frac{1}{t}(W(t)x - W(0)x) = W_1(t)\left[\frac{1}{t}(W_2(t)x - W_2(0)x)\right]$$

$$+ W_2(0)\left[\frac{1}{t}(W_1(t)x - W_1(0)x)\right].$$

Since $\|W_1(t)\|$ is bounded, for t small, and x is in $\mathcal{D}(B_1) \cap \mathcal{D}(B_2)$, we have

$$\lim_{t\to 0}\frac{1}{t}(W(t)x - W(0)x) = W_1(0)(W_2(0)B_2x) + W_2(0)(W_1(0)B_1x)$$

$$= W(0)Ax.$$

Thus an extension of A generates $W(t)$.

Since $W_1(t)$ and $W_2(t)$ commute, $W(t)$ commutes with all resolvents of B_1 and B_2. This implies that $W(t)$ leaves $\mathcal{D}(A)$ invariant. ∎

Example 15.3. We define Φ, on $B(X)$, by

$$\Phi(B) \equiv A_1B + BA_2.$$

When A_1 and A_2 generate strongly continuous semigroups, then, formally, $e^{t\Phi}(B) = e^{tA_1}Be^{tA_2}$. Since e^{tA_1} and e^{tA_2} are only strongly continuous and not norm continuous, $e^{t\Phi}$ will not be strongly continuous. However, $(r - A_1)^{-1}e^{tA_1}$ and $e^{tA_2}(r - A_2)^{-1}$ *are* norm continuous, thus

$$W(t)(B) \equiv (r - A_1)^{-1}\left(e^{t\Phi}(B)\right)(r - A_2)^{-1}$$

is strongly continuous.

Thus, when A_1 and A_2 generate strongly continuous semigroups, we expect Φ to generate a Λ-regularized semigroup, where $\Lambda(B) \equiv (r - A_1)^{-1}B(r - A_2)^{-1}$. By a happy coincidence, the range of Λ is a comfortable domain for Φ.

Theorem. *Suppose A_1 and A_2 generate exponentially bounded n-times integrated semigroups, and r is in $\rho(A_1) \cap \rho(A_2)$.*

Define Λ, Φ, on $B(X)$, by $\Lambda(B) \equiv (r - A_1)^{-1}B(r - A_2)^{-1}$, $\Phi(B) \equiv A_1B + BA_2$, $\mathcal{D}(\Phi) \equiv$ range of Λ.

95

Then an extension of Φ generates a $\Lambda^{(n+1)}$-regularized semigroup, that leaves $\mathcal{D}(\Phi)$ invariant.

Proof: For $i = 1, 2$, let $S_i(t)$ be the $(r - A_i)^{-n}$-regularized semigroup generated by A_i (as guaranteed by Theorem 18.3).

Define $W(t)$, on $B(X)$, by $(W(t))(B) \equiv (r - A_1)^{-1}S_1(t)BS_2(t)(r - A_2)^{-1}$. Using the mean value theorem and the fact that $\frac{d}{dt}S_i(t)(r-A_i)^{-1}x$ exists and equals $S_i(t)A_i(r - A_i)^{-1}x$, for all x in X, $i = 1, 2$, it is not difficult to show that $(r - A_i)^{-1}S_i(t)$ is a continuous function of t, in the operator norm of $B(X)$. Thus $W(t)$ is a strongly continuous function of t.

It is clear that $W(0)$ equals Λ^{n+1}. To see that $W(t)$ is a Λ^{n+1}-regularized semigroup, we calculate

$$W(t)W(s)(B) = (r - A_1)^{-1}S_1(t)\left[(r - A_1)^{-1}S_1(s)BS_2(s)(r - A_2)^{-1}\right] \cdot$$

$$\cdot S_2(t)(r - A_2)^{-1}$$

$$= (r - A_1)^{-2}\left[(r - A_1)^{-n}S_1(t + s)BS_2(t + s)(r - A_2)^{-n}\right] \cdot$$

$$\cdot (r - A_2)^{-2},$$

$$= \left[\Lambda^{n+1}W(s + t)\right](B),$$

as desired.

Suppose now that B is in $\mathcal{D}(\Phi)$. Then

$$W(t)(B) = (r - A_1)^{-2}S_1(t)\left(\Lambda^{-1}(B)\right)S_2(t)(r - A_2)^{-2}.$$

Again using the mean value theorem, one may show that $(r-A_i)^{-2}S_i(t)$ is a differentiable function of t, in the operator norm of $B(X)$, for $i = 1, 2$.

Thus $W(t)(B)$ is differentiable, with

$$\frac{d}{dt}W(t)(B) = \left[\frac{d}{dt}(r - A_1)^{-2}S_1(t)\right]\Lambda^{-1}(B)S_2(t)(r - A_2)^{-2}$$

$$+ (r - A_1)^{-2}S_1(t)\Lambda^{-1}(B)\left[\frac{d}{dt}S_2(t)(r - A_2)^{-2}\right]$$

$$= (r - A_1)^{-1}S_1(t)A_1(r - A_1)^{-1}\Lambda^{-1}(B)(r - A_2)^{-1}S_2(t)(r - A_2)^{-1}$$

$$+ (r - A_1)^{-1}S_1(t)(r - A_1)^{-1}\Lambda^{-1}(B)(r - A_2)^{-1}A_2S_2(t)(r - A_2)^{-1}$$

$$= (r - A_1)^{-1}S_1(t)(A_1B)S_2(t)(r - A_2)^{-1}$$

$$+ (r - A_1)^{-1}S_1(t)(BA_2)S_2(t)(r - A_2)^{-1}$$

$$= W(t)\Phi(B).$$

This implies that an extension of Φ generates $W(t)$. Since $W(t)$ commutes with Λ, $W(t)$ leaves $\mathcal{D}(\Phi)$ invariant. $\blacksquare$

XVI. EXISTENCE AND UNIQUENESS FAMILIES

Let us recall the difference between a C-existence family, or mild C-existence family, and a C-regularized semigroup. We think of all these families as being $e^{tA}C$; for a C-regularized semigroup, $e^{tA}C = Ce^{tA}$, and this leads to two advantages: an algebraic definition of a C-regularized family, and not only existence, but uniqueness, of solutions of the abstract Cauchy problem (0.1).

A close look at the proofs of uniqueness, when A generates a C-regularized semigroup, reveals that what is being used is Ce^{tA}. If we drop the hypothesis that C commutes with A, then we could have two families of operators, $e^{tA}C$ (for existence) and Ce^{tA} (for uniqueness). More generally, we could have bounded operators C_1, C_2, and consider families of operators $e^{tA}C_1$, C_2e^{tA}. It is only C_2 that need be injective, since $e^{tA}C_1$ is merely providing existence of solutions, for all initial data in the image of C_1. The family of operators C_2e^{tA}, besides providing uniqueness, also suggests the possibility of defining a *generator*, as with C-regularized semigroups: $A \equiv C_2^{-1}(\frac{d}{dt}C_2e^{tA})|_{t=0}$. Finally, since

$$(C_2e^{tA})(e^{sA}C_1) = C_2e^{(t+s)A}C_1 = C_2(e^{(t+s)A}C_1) = (C_2e^{(t+s)A})C_1,$$

we obtain an algebraic definition of the *pair* of families of operators $(e^{tA}C_1, C_2e^{tA})$ (Definition 16.1).

To summarize: existence and uniqueness have been separated into two families of operators, that are equal when C_1 and C_2 are equal and commute with A.

To define the algebraic properties of the (C_1, C_2) existence and uniqueness family $\{(W_1(t), W_2(t))\}_{t \geq 0}$, it is necessary to intertwine W_1 and W_2, as in Definition 16.1 (3) (see Remark 16.2). The basic properties of the generator are in Theorem 16.5. The relationships with the abstract Cauchy problem are in Theorem 16.5 and Propositions 16.6, 16.7 and 16.8.

Definition 16.1. The pair $\{(W_1(t), W_2(t)\}_{t \geq 0}$ of strongly-continuous families of bounded operators is a *mild (C_1, C_2) existence and uniqueness family* if

(1) $W_i(0) = C_i$, for $i = 1, 2$;

(2) C_2 is injective; and

(3) $W_2(t)W_1(s) = C_2W_1(t+s) = W_2(t+s)C_1$, for all $s, t \geq 0$.

The operator A generates (W_1, W_2) if

$$Ax = C_2^{-1} \left(\frac{d}{dt} W_2(t)x|_{t=0} \right),$$

$$\mathcal{D}(A) \equiv \{x \mid \frac{d}{dt} W_2(t)x \text{ exists, and equals } W_2(t) C_2^{-1} \left(\frac{d}{dt} W_2(t)x|_{t=0} \right),$$

$$\text{for all } t \geq 0\}.$$

Intuitively, $W_1(t) = e^{tA}C_1$, $W_2(t) = C_2 e^{tA}$.

Remark 16.2. When C_1 equals C_2, and commutes with $W_1(t)$ and $W_2(t)$, for all $t \geq 0$, then $W_1(t)$ equals $W_2(t)$, and $W_1(t)$ is a C_1-regularized semigroup. A consequence of Theorem 3.4(d) is that A generates $W_1(t)$.

First, we would like to give a simple example of an operator that generates a (C_1, C_2) existence and uniqueness family, but does not generate a regularized semigroup.

Example 16.3. Let $X \equiv \{\text{continuous } f : \mathbf{R} \to \mathbf{C} \mid \lim_{|x|\to\infty} f(x)e^{x^2} = 0\}$, and let $\|f\| \equiv \sup_{x \in \mathbf{R}} |f(x)e^{x^2}|$, $A \equiv d/dx$, with maximal domain.

Let $(C_1 f)(x) \equiv e^{-x^2} f(x)$, $C_2 \equiv C_1$,

$$(W_1(t)f)(x) \equiv e^{-(x+t)^2} f(x+t)$$

$$(W_2(t)f)(x) \equiv e^{-x^2} f(x+t).$$

(Intuitively, $W_1(t) = e^{tA}C_1$, $W_2(t) = C_2 e^{tA}$, where e^{tA} is translation, $(e^{tA}f)(x) \equiv f(x+t)$.)

Then it is straightforward to show that (W_1, W_2) is a (C_1, C_2) existence and uniqueness family, generated by A. In fact, the abstract Cauchy problem has a unique solution, for all initial data in a dense set(see Theorem 16.5(b)). However, the following demonstrates that regularized semigroups, unlike existence families, are inadequate to produce these solutions.

Proposition 16.4. *If A is as in Example 16.3, then the abstract Cauchy problem (0.1) has a unique solution, for all initial data in a dense set, but A does not generate a C-regularized semigroup, for any C.*

Theorem 16.5. *Suppose A generates the mild (C_1, C_2) existence and uniqueness family (W_1, W_2). Then*

(a) *A is closed;*

(b) *$\{W_1(t)\}_{t\geq 0}$ is a mild C_1-existence family for A; and*

(c) *$\overline{\mathcal{D}(A)} \supseteq \overline{Im(C_1)}$.*

98

Proposition 16.6. *Suppose A generates a mild (C_1, C_2) existence and uniqueness family. Then all solutions and mild solutions of the abstract Cauchy problem (0.1) are unique.*

As with regularized semigroups (see Theorem 3.7), we may obtain partial converses of Theorem 16.5(b).

As long as $\mathcal{D}(A)$ is sufficiently large, it is sufficient that an extension of A be such a generator.

Proposition 16.7. *Suppose $\{W_1(t)\}_{t\geq 0}$ and $\{W_2(t)\}_{t\geq 0}$ are strongly continuous families of bounded operators, and $\int_0^t W_1(s)x \, ds \in \mathcal{D}(A)$, for all $x \in E$. Then the following are equivalent.*

(a) *(W_1, W_2) is a mild (C_1, C_2) existence and uniqueness family, generated by an extension of A.*

(b) *$\{W_1(t)\}_{t\geq 0}$ is a mild C_1-existence family for A, $W_2(0) = C_2$, and $\frac{d}{dt}W_2(t)x$ exists and equals $W_2(t)Ax$, for all $x \in \mathcal{D}(A)$, $t \geq 0$.*

Similar results for strong solutions are in the following.

Proposition 16.8. *Suppose $\{W_1(t)\}_{t\geq 0}$ and $\{W_2(t)\}_{t\geq 0}$ are strongly continuous families of bounded operators, $\mathcal{D}(A)$ is dense, and $t \mapsto W_1(t)x$ is a continuous map from $[0,\infty)$ into $[\mathcal{D}(A)]$, for all $x \in \mathcal{D}(A)$. Then the following are equivalent.*

(a) *(W_1, W_2) is a (C_1, C_2) existence and uniqueness family, generated by an extension of A.*

(b) *$\{W_1(t)|_{[\mathcal{D}(A)]}\}_{t\geq 0}$ is a C_1-existence family for A, $W_2(0) = C_2$, and $\frac{d}{dt}W_2(t)x$ exists and equals $W_2(t)Ax$, for all $x \in \mathcal{D}(A)$, $t \geq 0$.*

Example 16.9. For $n \in \mathbf{N}$, define, on $X_1 \times X_2$,

$$A = \begin{pmatrix} G_1 & B \\ 0 & G_2 \end{pmatrix}, \qquad \mathcal{D}(A) \equiv \mathcal{D}(G_1) \times [\mathcal{D}(B) \cap \mathcal{D}(G_2)],$$

where $\mathcal{D}(G_2^n) \subseteq \mathcal{D}(B)$, B is closed, and G_i generates a strongly-continuous semigroup, for $i = 1, 2$ and there exist $m \in \mathbf{N}$, $w \in \rho(G_1)$ such that $(w - G_1)^{-m}B$ is bounded.

Then an extension of A generates a (C_1, C_2) existence and uniqueness family, where

$$C_1 \equiv \begin{bmatrix} I & 0 \\ 0 & (s - G_2)^{-n} \end{bmatrix}, C_2 \equiv \begin{bmatrix} (w - G_1)^{-m} & 0 \\ 0 & I \end{bmatrix}.$$

Proposition 16.4 is shown by the following two lemmas.

Lemma 16.10. *Suppose $W(t)$ is a C-regularized semigroup generated by A, from Example 16.3. Then there exists μ, a complex-valued measure of bounded variation on bounded intervals, such that, for all f in $C_c^\infty(\mathbf{R})$,*

$$(Cf)(x) = \int_{\mathbf{R}} f(x+r)\, d\mu(r)$$

$$(W(t)f)(x) = \int_{\mathbf{R}} f(x+t+r)\, d\mu(r).$$

Proof: Fix $f \in C_c^\infty(\mathbf{R})$. Since the generator of a C-regularized semigroup commutes with C (see Theorem 3.4), $Cf \in \mathcal{D}(A)$, and $\frac{d}{dx}(Cf) = C(\frac{df}{dx})$. Let $g_t(x) \equiv g(x+t)$, for x real. Since C is bounded and $f \in C_c^\infty(\mathbf{R})$, $\lim_{h\to 0} \frac{1}{h}(f_{t+h} - f_t)$ exists, in X, and equals f_t', for all $t \geq 0$. Thus, since C is bounded, $\frac{d}{dt}C(f_t) = \frac{d}{dx}C(f_t)$. Since $\frac{d}{dx}$, on $C_0(\mathbf{R})$, with maximal domain, generates a strongly continuous semigroup, and $\frac{d}{dt}(Cf)_t = \frac{d}{dx}(Cf)_t$ (in $C_0(\mathbf{R})$), it follows that

$$C(f_t) = (Cf)_t,$$

for all $t \geq 0$.

For all real x, there exists μ_x, a complex-valued measure of bounded variation on bounded intervals, such that

$$(Cg)(x) = \int_{\mathbf{R}} g(r)\, d\mu_x(r),$$

for all $g \in X$. For $f \in C_c^\infty(\mathbf{R})$, $x \in \mathbf{R}$,

$$\int_{\mathbf{R}} f(r)\, d\mu_x(r+x) = \int_{\mathbf{R}} f(w-x)\, d\mu_x(w) = [C(f_{-x})](x) = (Cf)_{-x}(x)$$

$$= (Cf)(0) = \int_{\mathbf{R}} f(r)\, d\mu_0(r).$$

Thus $d\mu_x(r+x) = d\mu_0(r)$, for all real r, x, so that

$$(Cf)(x) = \int_{\mathbf{R}} f(w)\, d\mu_0(w-x)$$

$$= \int_{\mathbf{R}} f(r+x)\, d\mu_0(r),$$

the desired representation of C.

Since $\frac{d}{dt}\int_{\mathbf{R}} f(x+t+r)\, d\mu(r) = \frac{d}{dx}\int_{\mathbf{R}} f(x+t+r)\, d\mu(r)$, for any f in $C_c^\infty(\mathbf{R})$, $W(t)$ is as stated. $\blacksquare$

The following lemma shows that we cannot have both C and $W(t)$ in $B(X)$, for any $t > 0$.

Lemma 16.11. *Suppose $t \geq 0$, and $W(t)$, as in Lemma 16.10, is in $B(X)$. Then the support of μ is contained in $\{-t\}$.*

Proof: For any real x, the map $f \mapsto e^{x^2}(W(t)f)(x)$ is a linear functional on X, of norm less than or equal to $\|W(t)\|$. Thus there exists a complex-valued measure, ψ_x, of total variation $|\psi_x| \leq \|W(t)\|$, such that

$$e^{x^2} \int_{\mathbf{R}} f(x + t + r)\, d\mu(r) = \int_{\mathbf{R}} e^{w^2} f(w)\, d\psi_x(w),$$

for all $f \in C_c^\infty(\mathbf{R})$, which implies that

$$\|W(t)\| \geq |\psi_x| = \int_{\mathbf{R}} e^{x^2 - w^2}\, d|\mu|(w - x - t)$$

$$= \int_{\mathbf{R}} e^{x^2} e^{-(x+t+r)^2}\, d|\mu|(r)$$

$$= \int_{\mathbf{R}} e^{-2x(t+r)} e^{-(t+r)^2}\, d|\mu|(r)$$

$$= \int_{\mathbf{R}} e^{xw} e^{-w^2/4}\, d|\mu|(-t - \frac{w}{2}).$$

A short argument with Liouville's theorem shows that, because this is a bounded function of $x \in \mathbf{R}$, the support of $|\mu|$ must be contained in $\{-t\}$. ∎

Proof of Theorem 16.5: (a) It's clear from the definition that $\frac{d}{dt}W_2(t)x$

$= W_2(t)Ax$, for all $x \in \mathcal{D}(A)$, $t \geq 0$. Suppose $\{x_n\}_{n=1}^\infty \subseteq \mathcal{D}(A)$, $x_n \to x$ and $Ax_n \to y$, as $n \to \infty$. Then $W_2(t)x_n$ converges to $W_2(t)x$, and $\frac{d}{dt}W_2(t)x_n = W_2(t)Ax_n$ converges to $W_2(t)y$, both uniformly on compact subsets of $[0, \infty)$. The strong continuity of $W_2(t)$ now implies that $\frac{d}{dt}W_2(t)x$ exists, and equals $W_2(t)y$, for all $t \geq 0$. This implies that $x \in \mathcal{D}(A)$, and $Ax = y$, as desired.

(b) For $x \in X$, $t > 0$, let

$$y \equiv \int_0^t W_1(s)x\, ds.$$

For any $r, r + h > 0$,

$$\frac{1}{h}\left(W_2(r+h)y - W_2(r)y\right) = \frac{1}{h}\left[\int_0^t W_2(r)W_1(s+h)x - W_2(r)W_1(s)x\,ds\right]$$

$$= \frac{1}{h}W_2(r)\left(\int_h^{t+h} - \int_0^t\right)(W_1(s)x\,ds)$$

$$= W_2(r)\left[\frac{1}{h}\left(\int_t^{t+h} - \int_0^h\right)(W_1(s)x\,ds)\right],$$

which converges to $W_2(r)(W_1(t)x - C_1x)$, as $h \to 0$. Thus $y \in \mathcal{D}(A)$, with $Ay = W_1(t)x - C_1x$, as desired.

Assertion (c) follows from (b), since $C_1x = \lim_{h\to 0}\frac{1}{h}\int_0^h W_1(s)x\,ds$, for any $x \in E$. $\blacksquare$

Proof of Proposition 16.6: Suppose u is a solution or mild solution of (0.1), with $x = 0$. Then, for all $t \geq s \geq 0$, $\frac{d}{ds}[W_2(t-s)u(s)] = 0$, so that $C_2u(t) = W_2(t)u(0) = 0$. Since C_2 is injective, this implies that $u(t) = 0$, for all $t \geq 0$, as desired. $\blacksquare$

Proof of Proposition 16.7: (a) $\to$ (b) is immediate from Theorem 16.5(b) and the definition of the generator.

(b) $\to$ (a). For $t \geq w \geq 0$, $x \in X$,

$$\int_w^t W_2(r)C_1x\,dr = \int_0^{t-w} W_2(t-s)C_1x\,ds$$

$$= \int_0^{t-w}\frac{d}{ds}\left[W_2(t-s)\left(\int_0^s W_1(r)x\,dr\right)\right]ds$$

$$= W_2(w)\left(\int_0^{t-w} W_1(r)x\,dr\right), \quad \text{so that}$$

$$W_2(t)C_1x = W_2(w)W_1(t-w)x.$$

$\blacksquare$

Proof of Proposition 16.8: (a) $\to$ (b). Let $\tilde{A}$ be the generator of (W_1, W_2), $x \in \mathcal{D}(A)$. By Theorem 16.5(b),

$$W_1(t)x = C_1x + \tilde{A}\left(\int_0^t W_1(s)x\,ds\right)$$

$$= C_1x + \int_0^t AW_1(s)x\,ds,$$

102

since $\tilde{A}$ is closed (by Theorem 16.5(a)), and by hypothesis, $W_1(s)x \in \mathcal{D}(A)$ and $s \mapsto AW_1(s)x$ is continuous.

The derivative formula for W_2 follows immediately from the definition of the generator.

(b) → (a). For $x \in \mathcal{D}(A)$, $0 = \frac{d}{ds} W_2(t-s)W_1(s)x$, for all $t \geq s \geq 0$, thus $W_2(0)W_1(t)x = W_2(t-s)W_1(s)x = W_2(t)W_1(0)x$. Since $\mathcal{D}(A)$ is dense and $W_i(t) \in B(X)$, for $i = 1, 2$, $t \geq 0$, the same is true for all $x \in X$. Since $\frac{d}{dt} W_2(t)x = W_2(t)Ax$, for all $x \in \mathcal{D}(A)$, the generator of (W_1, W_2) is an extension of A. ∎

Proof of Example 16.9: As in Example 2.11, formally

$$e^{tA} = \begin{pmatrix} e^{tG_1} & \int_0^t e^{(t-w)G_1} B e^{wG_2}\, dw \\ 0 & e^{tG_2} \end{pmatrix},$$

and we want $W_1(t) = e^{tA}C_1, W_2(t) = C_2 e^{tA}$; thus our choice of C_1 and C_2 become clear, as we make the following definitions.

For $t \geq 0$, let

$$W_1(t) \equiv \begin{pmatrix} e^{tG_1} & \int_0^t e^{(t-w)G_1} B(s - G_2)^{-n} e^{wG_2}\, dw \\ 0 & e^{tG_2}(s - G_2)^{-n} \end{pmatrix}.$$

For the uniqueness family, define, for x_2 in $\mathcal{D}(G_2^n)$, $t \geq 0$, $x = (x_1, x_2)$,

$$W_2(t)x \equiv \begin{pmatrix} (w - G_1)^{-m} e^{tG_1} & \int_0^t e^{(t-y)G_1} (w - G_1)^{-m} B e^{yG_2}\, dy \\ 0 & e^{tG_2} \end{pmatrix} x.$$

Since G_2 generates a strongly continuous semigroup, $\mathcal{D}(G_2^n)$ is dense. Our hypotheses imply that $W_2(t)$ is bounded. Thus $W_2(t)$ may be extended to a strongly continuous family of bounded operators on $X_1 \times X_2$.

A calculation shows that $W_2(t)$ and $W_1(s)$ satisfy (3) of Definition 16.1, for $x_2 \in \mathcal{D}(G_2^n)$, and hence for all x, since all relevant operators are bounded. Thus $\{W_1(t), W_2(t)\}_{t \geq 0}$ is a (C_1, C_2) existence and uniqueness family. To see that A is the generator, note that, for $x \in \mathcal{D}(A)$, the map $t \mapsto W_2(t)x$ is differentiable, with

$$\frac{d}{dt} W_2(t)x|_{t=0} = \begin{bmatrix} G_1(w - G_1)^{-m} & (w - G_1)^{-m} B \\ 0 & G_2 \end{bmatrix} x,$$

which is in the image of C_2. A calculation shows that, for $x \in \mathcal{D}(A)$, $\frac{d}{dt} W_2(t)x$ exists and equals $W_2(t)Ax$. Thus an extension of A is the generator. ∎

XVII. *C*-RESOLVENTS AND HILLE-YOSIDA TYPE THEOREMS

Theorems about semigroups that involve resolvents are desirable, because resolvents are often relatively easy to calculate or estimate. For *C*-regularized semigroups or *C*-existence families, we need a generalization of resolvent that we introduced in Definition 3.6, the *C-resolvent*.

Example 17.1. Except when *C* is the identity, there is no reason to expect the generator of a *C*-regularized semigroup, even a bounded regularized semigroup, to have nonempty resolvent set. Consider

$$A \equiv \begin{bmatrix} G & G^k \\ 0 & G \end{bmatrix}, \; \mathcal{D}(A) \equiv \mathcal{D}(G) \times \mathcal{D}(G^k),$$

where $k \in \mathbf{N}$ and G is unbounded and generates a strongly continuous exponentially decaying semigroup (so that $0 \in \rho(G)$). Then an extension of A generates the bounded G^{-k}-regularized semigroup

$$W(t) \equiv \begin{bmatrix} e^{tG}G^{-k} & te^{tG} \\ 0 & e^{tG}G^{-k} \end{bmatrix}.$$

However, for $k > 2$, it is not hard to show that A, and any extension of A, has empty resolvent set. If $(\lambda - G)$ fails to be injective (surjective), then the same is true of $(\lambda - A)$, while if $\lambda \in \rho(G)$, then a calculation shows that, at least on a dense set,

$$(\lambda - A)^{-1} = \begin{bmatrix} (\lambda - G)^{-1} & G^k(\lambda - G)^{-2} \\ 0 & (\lambda - G)^{-1} \end{bmatrix}.$$

For $k > 2$, this implies that $(\lambda - A)^{-1}$ is an unbounded operator, thus all extensions of A, including the generator of $\{W(t)\}_{t \geq 0}$, have nonempty resolvent set.

The following proposition shows that the natural analogue of resolvent, for generators of *C*-regularized semigroups, is the *C-resolvent*, introduced in Definition 3.6.

Proposition 17.2. *Suppose* $\omega \in \mathbf{R}$ *and* A *generates a* $O(e^{\omega t})$ *C-regularized semigroup. Then* $\{z \mid Re(z) > \omega\} \subseteq \rho_C(A)$.

The map $z \mapsto (z - A)^{-1}C$ *is a holomorphic map from* $Re(z) > \omega$ *into* $B(X)$.

Hille-Yosida type theorems are a consequence of our "pointwise Hille-Yosida theorems" in Chapter V.

104

Theorem 17.3. *Suppose* $\omega \in \mathbf{R}$, *A is closed,* $C \in B(X)$, $(\omega, \infty) \subseteq \rho_C(A)$ *and* $Im(C) \subseteq Im(s-A)^n$, *for all* $s > \omega, n \in \mathbf{N}$, *with*

$$\{\|(s-\omega)^n(s-A)^{-n}C\| \mid s > w, n \in \mathbf{N}\} \text{ bounded.}$$

Then for all $r > \omega$, *there exists a mild* $(r-A)^{-1}C$-*existence family for A,* $\{W(t)\}_{t\geq 0}$, *such that* $e^{-\omega t}W(t)$ *is Lipschitz continuous.*

One surprising consequence of the following is that, when A is closed and densely defined, $(\omega, \infty) \subseteq \rho_C(A)$, and $\{W(t)\}_{t\geq 0}$ is a C-regularized semigroup generated by an extension of A, that leaves $\mathcal{D}(A)$ invariant, then $\int_0^t W(s)x\, ds \in \mathcal{D}(A)$, for all $x \in X, t \geq 0$. We know that this is true when A itself is the generator (Theorem 3.4), but it seems surprising that the domain of A in this situation is sufficiently large; apparently having nonempty C-resolvent forces the domain of A to be large.

Note also that, in the following, if $Im(C)$ is dense, then $\mathcal{D}(A)$ is automatically dense (Theorem 3.4).

Theorem 17.4. *Suppose* $C \in B(X)$ *is injective,* $\omega \in \mathbf{R}$, *A is closed,* $\mathcal{D}(A)$ *is dense,* $(\omega, \infty) \subseteq \rho_C(A)$ *and* $CA \subseteq AC$. *Then the following are equivalent.*

(a) *An extension of A generates a* $O(e^{\omega t})$ *C-regularized semigroup that leaves* $\mathcal{D}(A)$ *invariant.*

(b) *There exists an* $O(e^{\omega t})$ *strong C-existence family,* $\{W(t)\}_{t\geq 0}$, *for A, such that* $W(t)A \subseteq AW(t)$, *for all* $t \geq 0$.

(c) $Im(C) \subseteq Im(s-A)^n$, *for all* $s > \omega, n \in \mathbf{N}$, *with*

$$\{\|(s-\omega)^n(s-A)^{-n}C\| \mid s > \omega, n \in \mathbf{N}\} \text{ bounded.}$$

(d) $Im(C) \subseteq Im(s-A)^n$, *for all* $s > \omega, n \in \mathbf{N}$, *with*

$$\{\|(s-\omega)^n(s-A)^{-n}x\| \mid s > \omega, n \in \mathbf{N}\} \text{ bounded,} \forall x \in Im(C).$$

Proposition 3.9 then immediately gives us the following.

Corollary 17.5. *Suppose* $C \in B(X)$ *is injective,* $\omega \in \mathbf{R}$, $\mathcal{D}(A)$ *is dense,* $(\omega, \infty) \subseteq \rho(A)$ *and* $CA \subseteq AC$. *Then the following are equivalent.*

(a) *A generates an* $O(e^{\omega t})$ *C-regularized semigroup.*

(b) *There exists an* $O(e^{\omega t})$ *strong C-existence family,* $\{W(t)\}_{t\geq 0}$, *for A, such that* $W(t)A \subseteq AW(t)$, *for all* $t \geq 0$.

(c) $\{\|(s-\omega)^n(s-A)^{-n}C\| \mid s > \omega, n \in \mathbf{N}\}$ *is bounded.*

And as an immediate corollary, using Theorem 3.4(a) and (b) and Proposition 17.2, we obtain the famous Hille-Yosida-Phillips-Miyadera theorem for strongly continuous semigroups.

105

Corollary 17.6 (Hille-Yosida-Phillips-Miyadera). *Suppose $\omega \in \mathbf{R}$. Then the following are equivalent.*

(a) *A generates a $O(e^{\omega t})$ strongly continuous semigroup.*

(b) *$\mathcal{D}(A)$ is dense, $(\omega, \infty) \subseteq \rho(A)$ and*

$$\{\|(s - \omega)^n (s - A)^{-n}\| \mid s > \omega, n \in \mathbf{N}\} \text{ is bounded.}$$

The following is a particularly simple sufficient condition for the existence of a mild existence family, since it involves only polynomial growth of the resolvent (not its powers, as in the previous theorems) in a half-plane.

Theorem 17.7. *Suppose A is closed, $\delta > 0$, $k \in \mathbf{N}$ and whenever $Re(z) > \delta, (z - A)$ is injective, $Im(C) \subseteq Im(z - A)^{k+1}$ and the map $z \mapsto (z - A)^{-1}C$ is holomorphic and $O(|z|^{k-1})$ on $Re(z) > \delta$.*

Then, for all $r, b > \delta$, there exists a mild uniformly continuous $O(e^{bt})$ $(A - r)^{-(k+1)}C$-existence family for A.

Corollary 17.8. *Suppose $\omega, k \in \mathbf{N}, \{z \mid Re(z) > \omega\} \subseteq \rho(A)$ and*

$$\{|z|^{1-k}\|(z - A)^{-1}C\| \mid Re(z) > \omega\}$$

is bounded.

Then there exists a mild uniformly continuous exponentially bounded $(A - r)^{-(k+1)}C$-existence family for A, for all $r > \omega$.

Example 17.9. Theorem 17.7 may be applied to the operator considered in Example 16.9, to conclude the following.

Proposition 17.10. *If A and C_1 are as in Example 16.9, then there exists a mild uniformly continuous exponentially bounded $(A - r)^{-2}C_1$-existence family for A, for all $r > \omega$.*

This is in most ways a weaker result than appears in Example 16.9. The advantage of this technique is that resolvents are sometimes easier to calculate than semigroups, especially for larger matrices.

Let us write

$$\begin{pmatrix} A_{11} & A_{12} & \cdots & A_{1n} \\ \vdots & \ddots & & \vdots \\ \vdots & & \ddots & \vdots \\ A_{n1} & \cdots & \cdots & A_{nn} \end{pmatrix},$$

acting on $\overset{n}{\underset{i=1}{\times}} X_1$, to mean that A_{ij} maps a subspace of X_j into X_i for $1 \le i, j \le n$.

In [dL9], we apply Theorem 17.7 to the following.

Suppose $k_1 + 1, \ldots, k_n + 1, N_2, \ldots, N_n, M_1, \ldots, M_{n-1} \in \mathbf{N}$,

$$A = \begin{pmatrix} G_1 & B_{1,2} & \cdots & B_{1,n} \\ 0 & G_2 & \ddots & \vdots \\ \vdots & \ddots & \ddots & B_{n-1,n} \\ 0 & \cdots & 0 & G_n \end{pmatrix}$$

$$\mathcal{D}(A) \equiv \mathcal{D}(G_1) \overset{n}{\underset{j=2}{\times}} \mathcal{D}(G_j^{N_j}), \quad \text{where}$$

(1) $\mathcal{D}\left(G_j^{N_j}\right) \subseteq \mathcal{D}(B_{i,j})$, for $1 \le i < j \le n$;

(2) there exists M such that $(r - G_i)^{-1}$ exists, and $\|(r - G_i)^{-1}\|$ is $O(|r|^{k_i})$, for $1 \le i \le n$, $Re(r) > M$;

(3) there exists $s \in \mathbf{C}$ such that $(s - G_i)^{N_i} B_{i,j}(s - G_j)^{-N_j}$ is in $B(X_j, X_i)$, for $1 \le i < j \le n$ ($N_1 \equiv 0$); and

(4) there exists $s \in \mathbf{C}$ such that $(s - G_i)^{-M_i} B_{i,j}(s - G_j)^{M_j}$ is bounded, for $1 < i < j \le n$.

The proof is similar to the proof of Proposition 17.10.

As one special case that might be of interest, we could choose $X_j = L^p(\mathbf{R}^N)$ ($1 \le p \le \infty$), G_j equal to a constant coefficient differential operator $p_j(D)$, where $D = (i(\partial/\partial x_1), \ldots, i(\partial/\partial x_N))$, p_j is an elliptic nonconstant polynomial such that $\{Re(p_j(x)) \,|\, x \in \mathbf{R}^N\}$ is bounded above and $B_{i,j}$ is equal to a linear partial differential operator, $B_{i,j} = \sum_{|\alpha| \le m_{i,j}} h_{\alpha,i,j} D^\alpha$, of arbitrary order, where $h_{\alpha,i,j}$ is infinitely differentiable, with bounded derivatives.

More precisely, let $\mathcal{D}(B_{i,j}) \equiv W^{m_{i,j},p}(\mathbf{R}^N)$, and

$$\mathcal{D}(G_j) \equiv \{f \in L^p(\mathbf{R}^N) | p_j(D)f \in L^p(\mathbf{R}^N) \text{ distributionally}\}.$$

To see that the theorem of Example 17.8 applies, note that, for all j, since p_j is elliptic and nonconstant, there exists P_j such that

$$\mathcal{D}\left(G_j^{P_j}\right) \subseteq W^{1,p}(\mathbf{R}^N).$$

We construct $N_1, \ldots, N_n$ inductively as follows. Let $N_1 \equiv 0$. If $N_1, \ldots, N_k$ are chosen that satisfy (3), let $N_{k+1} \equiv P_{k+1} \max\{N_i \text{ord}(p_i) + m_{i,k+1} | 1 \le i \le k\}$. Then, for all $i < k + 1$,

$$(s - G_{k+1})^{-N_{k+1}} : X_{k+1} \mapsto W^{(N_i \text{ord}(p_i) + m_{i,k+1}),p}(\mathbf{R}^N),$$

107

so that

$$B_{i,k+1}(s - G_{k+1})^{-N_{k+1}} : X_{k+1} \mapsto W^{N_i \text{ord}(p_i),p}(\mathbf{R}^N)$$

and

$$(s - G_i)^{N_i} B_{i,k+1}(s - G_{k+1})^{-N_{k+1}} \in B(X_{k+1}, X_i),$$

as desired, completing the induction, and establishing conditions (1) and (3).

Condition (4) is established similarly, letting $M_n \equiv 0$, and applying induction on k to $\{M_{n-k}\}_{k=0}^{n-1}$.

Proof of Proposition 17.2: Let $\{W(t)\}_{t\geq0}$ be the C-regularized semigroup generated by A.

By Lemma 2.10 and Theorem 3.5(1),

$$(z - A) \int_0^\infty e^{-zt} W(t)x \, dt = Cx, \tag{17.11}$$

for all $x \in X, Re(z) > \omega$. Thus it is sufficient to show that $(z - A)$ is injective, whenever $Re(z) > \omega$.

Suppose $(z - A)x = 0$, where $x \in \mathcal{D}(A)$ and $Re(z) > \omega$. Then by (17.11), since A is closed (Theorem 3.4),

$$0 = \int_0^\infty e^{-zt} W(t)(z - A)x \, dt = (z - A) \int_0^\infty e^{-zt} W(t)x \, dt = Cx.$$

Since C is injective, this implies that $x = 0$.

After applying $(z - A)^{-1}$ to both sides of (17.11), we conclude that $z \mapsto (z - A)^{-1}C$ is a holomorphic map from $Re(z) > \omega$ into $B(X)$. ∎

Proof of Theorem 17.3: We may assume, by translating if necessary, that $\omega = 0$. By Theorem 5.10, $Im(C) \subseteq Y$, the weak solution space (see Definition 5.2). Thus $Im((r - A)^{-1}C) \subseteq \mathcal{D}(A|_Y) \subseteq Z_0$, the Hille-Yosida space (see Definition 5.1), so that by Theorem 5.16, there exists a bounded mild $(r - A)^{-1}C$-existence family for $A, W(t) \equiv e^{tA|_{Z_0}}(r - A)^{-1}C$.

Since $Im((r - A)^{-1}C) \subseteq \mathcal{D}(A|_Y)$, Theorem 5.5(6) implies that, for all $x \in X, W(t)x = u(t, (r-A)^{-1}Cx)$, the unique solution of (0.1) with initial data $(r - A)^{-1}Cx$, is Lipschitz continuous. By the uniform boundedness principle, $\{W(t)\}_{t\geq0}$ is Lipschitz continuous. ∎

Proof of Theorem 17.4: We may assume that $\omega = 0$. Let Y be the weak solution solution space for A and let Z_0 be the Hille-Yosida space for A (see Definitions 5.1 and 5.2).

108

(d) → (b). By Theorem 5.10, $Im(C) \subseteq Y$, thus, as in the proof of Theorem 5.16, $[Im(C)] \hookrightarrow Y$. We claim that $Im(C)$ is contained in Z_0. Suppose $x \in X$. There exists $<x_n> \subseteq \mathcal{D}(A)$ such that $x_n \to x$, in X, thus $Cx_n \to Cx$, in $[Im(C)]$ and hence in Y. Since $CA \subseteq AC$, $Cx_n \in \mathcal{D}(A|_Y) \subseteq Z_0$, for all n, thus $Cx \in Z_0$, proving the claim.

Theorem 5.16 now implies that there exists a mild C-existence family for A, $W(t)x \equiv u(t, Cx)$. Since $CA \subseteq AC$, the uniqueness of the solutions of (0.1) implies that $W(t)A \subseteq AW(t)$, for all $t \geq 0$. By Corollary 2.8, $\{W(t)\}_{t \geq 0}$ is a strong C-existence family for A.

(b) → (c). By Theorem 5.16, $[Im(C)] \hookrightarrow Y$ (note that, since $\mathcal{D}(A)$ is dense, the C-existence family is strongly uniformly continuous). The expression for $\|x\|_Y$ in Theorem 5.10 now implies (c).

(b) ↔ (a) follows from Theorem 3.7(a) and Theorem 3.8(3). ∎

Proof of Theorem 17.7: By Theorem 5.15, $Im((A - r)^{-(k+1)}C) \subseteq Z_b$. Thus this follows from Theorem 5.16. ∎

Lemma 17.12. *Suppose B is an injective operator, from a subspace of X_1 into X_2, and for $i = 1, 2$, there exists injective $D_i \in B(E_i)$ such that $D_2 B$ and $D_1 B^{-1}$ are bounded. Then B is closable and $\overline{B}$ is injective.*

Proof: Suppose $x_n \to 0$ and $Bx_n \to y$. Then $(D_2 B)x_n \to 0$, and $D_2(Bx_n) \to D_2 y$, so that $D_2 y = 0$, thus, since D_2 is injective, $y = 0$. This means that B is closable. Suppose $\overline{B}x = 0$. Then there exists $\{x_n\} \subseteq \mathcal{D}(B)$ such that $Bx_n \to 0$ and $x_n \to x$, that is, $B^{-1}(Bx_n) \to x$. The same argument, as given above, implies that $x = 0$, so that $\overline{B}$ is injective. ∎

Proof of Proposition 17.10: For r in $\rho(G_1) \cap \rho(G_2)$, $(r - A)$ is injective, with

$$(r - A)^{-1} = \begin{pmatrix} (r - G_1)^{-1} & (r - G_1)^{-1}B(r - G_2)^{-1} \\ 0 & (r - G_2)^{-1} \end{pmatrix}.$$

Since $\mathcal{D}(G_2^n) \subseteq \mathcal{D}(B)$, and B is closed, $(r - A)^{-1} \begin{pmatrix} I & 0 \\ 0 & (s-G_2)^{-n} \end{pmatrix}$ is in $B(X_1 \times X_2)$. By Lemma 17.12, $(r - A)$ is closable, and $(r - \overline{A})$ is injective, for $r \in \rho(G_1) \cap \rho(G_2)$.

Thus $\|(z - A)^{-1}C\|$ is bounded on right half-planes. The image of $(z - A)^2$ contains $X_1 \times \mathcal{D}(G_2^{n-1})$, since $\mathcal{D}(G_2^n) \subseteq \mathcal{D}(B)$, thus $Im(C) \subseteq Im(z - A)^2$, for $Re(z)$ large.

Theorem 17.7 thus implies that there exists a mild exponentially bounded $(A - r)^{-2}C$-existence family for A. ∎

109

XVIII. RELATIONSHIP TO INTEGRATED SEMIGROUPS

Another way to smooth an unbounded semigroup is by integrating. The following definition presents the algebraic properties that one would obtain by integrating a semigroup n times, and requires only (strong) continuity, rather than n times (strong) continuous differentiability, as one would obtain by integrating a strongly continuous semigroup n times.

Definition 18.1. If $n \in \mathbb{N}$, an *n-times integrated semigroup* is a strongly continuous family of operators $\{S(t)\}_{t \geq 0}$ such that $S(0) = 0$ and

$$S(t)S(s)$$
$$= \frac{1}{(n-1)!} \left[\int_t^{s+t} (s+t-r)^{n-1} S(r) \, dr - \int_0^s (s+t-r)^{n-1} S(r) \, dr \right],$$

for all $s, t \geq 0$.

$\{S(t)\}_{t \geq 0}$ is *nondegenerate* if, whenever $S(t)x = 0$, for all $t \geq 0$, then x must equal 0.

The *generator* is defined by

$$\mathcal{D}(A) = \{x \mid \exists y \text{ such that } S(t)x = \frac{t^n}{n!} x + \int_0^t S(r)y \, dr \, \forall t \geq 0\}, \quad \text{with } Ax = y.$$

We will use Chapter IV to show that generating an n-times integrated semigroup (not necessarily exponentially bounded) corresponds to generating an $(r-A)^{-n}$-regularized semigroup, whenever $r \in \rho(A)$.

Here is a simple example of a once-integrated semigroup that is not exponentially bounded.

Example 18.2. Let $Af(z) \equiv zf(z)$, on $C_0(\gamma)$, where $\gamma \equiv \{x + ie^{x^2} \mid x \geq 0\}$, with maximal domain.

The operator A generates the once-integrated semigroup

$$S(t)f(z) \equiv \int_0^t e^{rz} \, dr = \frac{1}{z}(e^{tz} - 1)f(z).$$

Then $\|S(t)\|^2 = \sup_{x \geq 0} \frac{(e^{ix}-1)^2}{(x^2+e^{2x^2})}$; a calculation shows that $\|e^{-\frac{1}{4}t^2} S(t)\|$ is bounded below as $t \to \infty$.

Note that A generates the A^{-1}-regularized semigroup

$$W(t)f(z) \equiv \frac{1}{z}e^{tz}f(z).$$

With a slight modification, we obtain a space, X, and an operator, A, such that the abstract Cauchy problem (0.1) has a unique mild solution, for all $x \in \mathcal{D}(A)$, but no nontrivial solutions of (0.1) are exponentially bounded (see Example 4.10).

Theorem 18.3. *Suppose X is a Banach space, $r \in \rho(A)$ and $n \in \mathbf{N}$. Let Z be as in Definition 4.6. Then the following are equivalent.*

(a) *The abstract Cauchy problem (0.1) has a unique solution, for all $x \in \mathcal{D}(A^{n+1})$.*

(b) *The abstract Cauchy problem has a unique mild solution, for all $x \in \mathcal{D}(A^n)$.*

(c) *All solutions of the abstract Cauchy problem are unique and*

$$[\mathcal{D}(A^n)] \hookrightarrow Z.$$

(d) *The operator A generates an $(r - A)^{-n}$-regularized semigroup, $\{W(t)\}_{t \geq 0}$.*

(e) *The operator A generates an n-times integrated nondegenerate semigroup, $\{S(t)\}_{t \geq 0}$.*

We then have $W(t)x = (\frac{d}{dt})^n S(t)(r - A)^{-n}x$ and $S(t) = (r - A)^n J^n W(t)x$, for all $x \in X$, where $Jf(t) \equiv \int_0^t f(s)\,ds$.

Here is an example of how we may use Theorem 18.3 to produce Hille-Yosida type theorems for integrated semigroups.

Corollary 18.4. *Suppose $\mathcal{D}(A)$ is dense and $n + 1 \in \mathbf{N}$. Then the following are equivalent.*

(a) *The operator A generates an exponentially bounded n-times integrated semigroup.*

(b) *There exists $w \in \mathbf{R}$ such that $(w, \infty) \subseteq \rho(A)$ and*

$$\{\frac{(s - w)^{k+1}}{k!}(\frac{d}{ds})^k \left(\frac{1}{s^n}(s - A)^{-1} \right) \mid s > w, k + 1 \in \mathbf{N}\}$$

is bounded.

Proof of Theorem 18.3: The equivalence of (a) through (d) is in Theorem 4.15, with $C \equiv (r - A)^{-n}$, since $\mathcal{D}(A^n) = Im((r - A)^{-n})$. It is straightforward to show that (e) implies (a) (see [A2] or [Th]), since an induction argument, using the definition of $\mathcal{D}(A)$, shows that

$$S(t)x = \sum_{j=0}^{k-1} \frac{t^{(n-j)}}{(n - j)!} A^j x + J^k(S(t)A^k x), \qquad (18.5)$$

for $x \in \mathcal{D}(A^k), 0 \leq k \leq n$.

111

Given (a) through (d), let $\{T(t)\}_{t\geq 0}$ be as in Theorem 4.8. By Theorem 4.15, $W(t) = T(t)(r - A)^{-n}$. For any $x \in X, J^nW(t)x = J^nT(t)(r - A)^{-n}x$, thus is in $\mathcal{D}((A|_Z)^n)$, since $T(t)$ is a strongly continuous semigroup generated by $A|_Z$. Let $S(t)x \equiv (r - A)^n J^n W(t)x$, for all $x \in X, t \geq 0$.

By translating A, we may assume that $r = 0$.

A tedious calculation shows that $\{S(t)\}_{t\geq 0}$ is an n-times integrated semigroup.

Note that, for all $x \in X$,

$$\int_0^t S(r)x\,dr = A^n J^{n+1} T(t) A^{-n} x \in \mathcal{D}(A),$$

since, by the properties of strongly continuous semigroups, $J^{n+1}T(t)A^{-n}x$ is in $\mathcal{D}((A|_Z)^{n+1})$; similarly,

$$A\left(\int_0^t S(r)x\,dr\right) = A^n J^n (T(t) - I) A^{-n} x = S(t)x - \frac{t^n}{n!}x. \qquad (18.6)$$

Let $\tilde{A}$ be the generator of $\{S(t)\}_{t\geq 0}$. If $x \in \mathcal{D}(A)$, then by (18.6), $S(t)x - \frac{t^n}{n!}x = \int_0^t S(r)Ax\,dr$, thus $x \in \mathcal{D}(\tilde{A})$ and $\tilde{A}x = Ax$, that is, $A \subseteq \tilde{A}$. Conversely, if $x \in \mathcal{D}(\tilde{A})$, then by (18.6) and the definition of the generator,

$$\int_0^t S(r)A^{-n}x\,dr = \int_0^t S(r)A^{-(n+1)}\tilde{A}x\,dr,$$

for all $r \geq 0$, so that we may differentiate $n + 1$ times to obtain

$$A^{-n}x = A^{-(n+1)}\tilde{A}x,$$

since, by (18.5), $(\frac{d}{dt})^n S(t)|_{t=0}x = x$, for all $x \in \mathcal{D}((\tilde{A})^n)$.

Thus $A = \tilde{A}$, which proves (e). ∎

Proof of Corollary 18.4: Since, for any $r \in \mathbf{C}, (A - r)$ generates an exponentially bounded n-times integrated semigroup if and only if A does (this is a tedious calculation), we may assume, by translating A if necessary, that $w = 0$ and $0 \in \rho(A)$. By Theorem 18.3, A generates an n-times integrated semigroup if and only if A generates an A^{-n}-regularized semigroup. The resolvent identity shows that

$$(s - A)^{-1}A^{-n} = \sum_{i=0}^{n-1} \frac{1}{s^{i+1}} A^{i-n} + \frac{1}{s^n}(s - A)^{-1}, \forall n \in \mathbf{N}, s \geq 0.$$

Thus, since $(-1)^k k!(s-A)^{-(k+1)}A^{-n} = (\frac{d}{ds})^k (s-A)^{-1}A^{-n}$, this corollary follows from Corollary 17.5. ∎

XIX. PERTURBATIONS

In this chapter we consider both additive and multiplicative perturbations. We see how regularized semigroups arise naturally as bounded, commuting multiplicative perturbations of the generator of a bounded, strongly continuous group (Theorem 19.6). This theorem is best possible (see Examples 19.7–19.9). It is interesting that, even when the multiplicative perturbation is bounded and has real spectrum and numerical range, the perturbed operator may not be the generator of a strongly continuous group (see Example 19.8). Since $BA = A + (B - 1)A$, a bounded multiplicative perturbation is, in general, an unbounded additive perturbation. We give other examples of additive unbounded perturbations, that transform the generator of a C_1-regularized semigroup into the generator of a C_2-regularized semigroup, where C_2 may have smaller range than C_1.

All the arguments for perturbations of generators of strongly continuous semigroups may be mimicked to obtain predictable analogous results for C-existence families, by placing a C everywhere. These are not the results of interest, and, as we commented in the introduction, we would like to avoid such things in this book. In fact, most such arguments may be avoided by using the solution space to reduce such results to the case $C = I$. An example is the following, which we leave as an exercise, using Theorem 5.16 and the fact that a bounded perturbation of the generator of a strongly continuous semigroup on a Banach space is also the generator of a strongly continuous semigroup.

Exercise 19.1. *Suppose there exists an exponentially bounded mild C-existence family for A and $B \in B([Im(C)], X)$. Then there exists an exponentially bounded mild C-existence family for $(A + B)$.*

Similar results may be obtained for existence and uniqueness families, with the mimicry we described above, but we do not feel these are worth mentioning. The hypothesis on the perturbing operator, B, in Exercise 19.1, is too limiting to be of much interest.

Of much more interest is to perturb A with an operator, B, that is *unbounded*, even relative to A. When there exists a C_1-existence family for A, we would like a C_2-existence family for $(A + B)$, where, in general, C_2 is more smoothing than C_1, that is, $C_1^{-1} C_2 \in B(X)$, but $C_2^{-1} C_1$ may not be. Two examples of this are Theorems 19.2 and 19.4.

Theorem 19.2. *Suppose*

(1) *A generates a bounded uniformly continuous C_1-regularized semigroup $\{W(t)\}_{t \geq 0}$ that commutes with C_2;*

113

(2) $C_1^{-1} C_2 \in B(X)$;

(3) B is closed in X and

(4) $W \equiv \{x \,|\, t \mapsto C_2^{-1} W(t) x$ is a bounded uniformly continuous map from $[0, \infty)$ into $X\} \subseteq \mathcal{D}(B)$, with $t \mapsto C_2^{-1} W(t) B x$ bounded and uniformly continuous, for all $x \in W$.

Then there exists an exponentially bounded mild C_2-existence family for $(A + B)|_W$.

Here is an example where B is unbounded relative to A.

Example 19.3. Suppose G generates a bounded strongly continuous semigroup, $\{T(t)\}_{t \geq 0}$, on X, B is closed in X, $r \in \rho(G)$ and $\mathcal{D}(G^n) \subseteq \mathcal{D}(B)$. Define A, on $X \times X$, by

$$A \equiv \begin{bmatrix} G & 0 \\ 0 & G \end{bmatrix}, \quad \mathcal{D}(A) \equiv \mathcal{D}(G) \times \mathcal{D}(G).$$

Let

$$C_2 \equiv \begin{bmatrix} I & 0 \\ 0 & (r - G)^{-n} \end{bmatrix} \in B(X \times X), \quad C_1 \equiv I, \quad W(t) \equiv \begin{bmatrix} T(t) & 0 \\ 0 & T(t) \end{bmatrix}.$$

Then Theorem 19.2 may be used to show that there exists a bounded mild C_2-existence family for

$$\tilde{A} \equiv \begin{bmatrix} G & B \\ 0 & G \end{bmatrix}, \quad \mathcal{D}(\tilde{A}) \equiv X \times \mathcal{D}(G^n),$$

since W, from (4), equals $X \times \mathcal{D}(G^n)$ and

$$\left\| C_2^{-1} W(T) \begin{bmatrix} 0 & B \\ 0 & 0 \end{bmatrix} \vec{x} \right\| = \|T(t) B x_2\| = \|T(t)[B(r - G)^{-n}](r - G)^n x_2\|,$$

whenever $x_2 \in \mathcal{D}(G^n)$.

Theorem 19.4 (Nilpotent perturbations). *Suppose there exist bounded, injective C_1, C_2 such that*

(1) A *generates an exponentially bounded C_1-regularized semigroup;*

(2) $BC_2 \in B(X)$;

(3) $\left[(r - A)^{-1} C_1 B C_2\right]^N = 0$, *for all $r \in \rho_{C_1}(A)$; and*

(4) C_2 *and C_1 commute with $BC_2, (r - A)^{-1} C_1, C_2$ and C_1, for all $r \in \rho_{C_1}(A)$.*

114

Then, for all $s \in \rho_{C_1}(A)$, there exists a mild exponentially bounded $(s-(A+B))^{-1}C_1(C_1C_2)^N$-existence family for $(A+B)$, where $\mathcal{D}(A+B) \equiv \mathcal{D}(A) \cap \mathcal{D}(B)$.

Example 19.5 Let G, A and $\tilde{A}$ be as in Example 19.3, except that $\mathcal{D}(\tilde{A}) \equiv \mathcal{D}(G) \times [\mathcal{D}(G) \times \mathcal{D}(B)]$, there exists $\lambda \in \rho(B)$ such that $(\lambda - B)^{-1}$ commutes with $T(t)$, for all $t \geq 0$, and we remove the hypothesis about the domain of G^n being contained in the domain of B.

Let $C \equiv (1 - \tilde{A})^{-1}(\lambda - B)^{-2}$.

Then Theorem 19.4 may be used to show that there exists a mild exponentially bounded C-existence family for $\tilde{A}$, by letting $C_1 \equiv I, C_2 \equiv (\lambda - B)^{-1}, N \equiv 2$, so that

$$(1-A)^{-1} \begin{bmatrix} 0 & B \\ 0 & 0 \end{bmatrix} C_2 = \begin{bmatrix} 0 & (1-G)^{-1}B(\lambda - B)^{-1} \\ 0 & 0 \end{bmatrix}.$$

This example may be extended to upper triangular matrices of arbitrary size, with commuting entries.

This example is weaker than 19.3, in that commuting is required, but is stronger in that there is no relationship between $\mathcal{D}(G)$ and $\mathcal{D}(B)$.

Our next theorem involves a *multiplicative* perturbation by a bounded operator. Note that, in general, this may be the same as an unbounded additive perturbation, since $AB = A + A(B - I)$.

Theorem 19.6. *Suppose $n \in \mathbb{N} \cup \{0\}$, B is bounded and A generates a bounded strongly continuous group that commutes with B. Then*

(a) *There exists bounded injective $C(\equiv e^{-A^2})$, with dense range, such that BA generates an entire C-regularized group;*

(b) *If $\{\|e^{itB}\|\}_{t \in \mathbb{R}}$ is $O(t^n)$, for t large, then BA generates an exponentially bounded $(1 + A)^{-(n+1)}$- regularized semigroup;*

(c) *If, in addition to (b) $0 \in \rho(B)$, then BA generates an exponentially bounded $(n + 1)$-times integrated semigroup.*

Example 19.7. This is a bounded operator B, and an operator A, as in Theorem 19.6, such that $\mathrm{sp}(B) = \{0\}$, but $\rho(BA) = \emptyset$. This shows that the invertibility of B is necessary in Lemma 19.12 and Theorem 19.6 (c) (at least for $n > 0$).

Let G be any unbounded generator of a bounded, strongly continuous group, on X, and, on $X \times X$, define

$$A \equiv \begin{bmatrix} G & 0 \\ 0 & G \end{bmatrix}, \quad B \equiv \begin{bmatrix} 0 & 1 \\ 0 & 0 \end{bmatrix}.$$

Then A generates a bounded strongly continuous group and $\|e^{tB}\| = \left\|\begin{bmatrix} 0 & t \\ 0 & 0 \end{bmatrix}\right\|$ is $O(t)$, but $BA = \begin{bmatrix} 0 & G \\ 0 & 0 \end{bmatrix}$ has empty resolvent, since $\text{Im}(r - BA) \subseteq X \times \mathcal{D}(G)$, for all complex r.

Example 19.8. Here we present B, A as in Theorem 19.6 (c), with $n = 0$, such that BA does not generate a bounded strongly continuous semigroup.

Let X be the set of all functions, f, from $\mathbf{R}^2$ into the complex plane, such that the map $y \mapsto f(x,y)$ is in $L^1(\mathbf{R})$, for all $x \in \mathbf{R}$ and the map $x \mapsto \int_{\mathbf{R}} |f(x,y)| \, dy$ is uniformly continuous and bounded, on $\mathbf{R}$, with norm

$$\|f\| = \sup\left\{ \int_{\mathbf{R}} |f(x,y)| \, dy \,\middle|\, x \in \mathbf{R} \right\}.$$

Let $h(y) = \sum_{k=1}^{\infty} \left(1 + (1/2k)\right) 1_{[k-1,k)}(y)$, and define the bounded operator B by

$$(Bf)(x,y) = h(y)f(x,y).$$

Let A be $\partial/\partial x$, the generator of $(e^{tA}f)(x,y) = f(x+t,y)$.

Note that B and e^{tA} commute, and the spectrum of B is contained in the closure of the range of h, which is contained in the interval $[1,2]$.

If e^{tBA} existed, it would be given by

$$(e^{tBA}f)(x,y) = f(x+th(y),y),$$

since

$$(d/dt)f(x+th(y),y) = h(y)(\partial f/\partial x)(x+th(y),y)$$
$$= (BAf)(x+th(y),y).$$

We will show that e^{BA} is unbounded. For any N, define f_N as follows. Let $A_N = \bigcup_{k=1}^{N}[1+(1/2k+1),1+(1/2k)] \times [k-1,k)$, $B_N = \bigcup_{k=1}^{N}[1+(1/2k),1+(1/2k-1)] \times [k-1,k)$. Then

$$f_N(x,y) \equiv \begin{cases} 1 & \text{for } x \text{ in } A_N \cap B_N \\ 0 & \text{for } x \text{ outside } A_N \cup B_N \quad \text{(see graphs below).} \\ \text{linear in } x & \text{on } A_N \text{ and } B_N \end{cases}$$

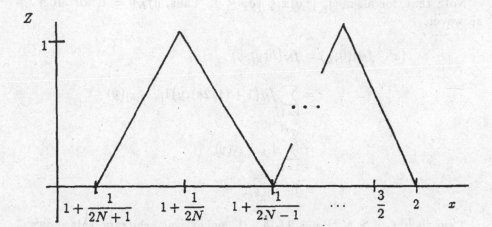

Figure 1. Projection of $z = f_N(x,y)$ onto x–z plane.

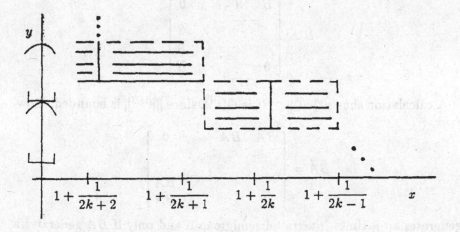

Figure 2. $\{(x,y) \mid f_N(x,y) \neq 0\}$

117

Note that, for all x, $\int_{\mathbf{R}} |f_N(x,y)|\, dy \leq 1$. Thus, $\|f_N\| = 1$, for all N. However,

$$(e^{BA} f_N)(0,y) = f_N(h(y),y)$$

$$= \sum_{k=1}^{\infty} f_N(1 + (1/2k),y) 1_{[k-1,k)}(y)$$

$$= \sum_{k=1}^{N} 1_{[k-1,k)}(y)$$

$$= 1_{[0,N)}(y).$$

Thus $\|e^{BA} f_N\| \geq N$. Since $\|f_N\| = 1$, and N was arbitrary, this shows that e^{BA} is unbounded.

Example 19.9. Let n be arbitrary. This is an example of $\tilde{B}$, $\tilde{A}$ as in Theorem 19.6 (b) and (c), such that $\tilde{B}\tilde{A}$ does not generate an $(1 + \tilde{A})^{-n}$-regularized semigroup or an n-times integrated semigroup. This shows that Theorem 19.6 (b) and (c) are best possible.

Let B, A be as in Example 19.8. On X^{n+1}, define

$$\tilde{A} \equiv \begin{bmatrix} A & & 0 \\ & \ddots & \\ 0 & & A \end{bmatrix},$$

$$\tilde{B} \equiv \begin{bmatrix} B & B & & 0 \\ & \ddots & \ddots & \\ & & \ddots & B \\ 0 & & & B \end{bmatrix}.$$

A calculation shows that $\|e^{it\tilde{B}}\|$ is $O(t^n)$, since $\|e^{itB}\|$ is bounded. However,

$$\tilde{B}\tilde{A} = \begin{bmatrix} BA & BA & & 0 \\ & \ddots & \ddots & \\ & & \ddots & BA \\ 0 & & & BA \end{bmatrix}$$

generates an n-times integrated semigroup if and only if BA generates a strongly continuous semigroup. Hence, by Example 19.8, $\tilde{B}\tilde{A}$ does not generate an n-times integrated semigroup. This implies that $\tilde{B}\tilde{A}$ does not generate a $(1 + \tilde{B}\tilde{A})^{-n}$-regularized semigroup (see Theorem 18.3);

118

since $(1 + \tilde{B}\tilde{A})^{-1}(1 + \tilde{A})$ is bounded and injective, $\tilde{B}\tilde{A}$ cannot generate a $(1 + \tilde{A})^{-n}$-regularized semigroup.

Example 19.10. For $1 \leq p < \infty$, $m \in \mathbf{N} \cup \{0\}$, let $X \equiv W^{m,p}(\mathbf{R}^2)$, the Sobolev space of all functions, f, in $L^p(\mathbf{R}^2)$, whose distributional derivative $D^\alpha f$, is in $L^p(\mathbf{R}^2)$, whenever $|\alpha| \leq m$, with

$$\|f\|_{m,p} \equiv \Big[\sum_{|\alpha| \leq m} \|D^\alpha f\|_p^p \Big]^{1/p}.$$

Let A be d/dx, the generator of the bounded strongly continuous group $(e^{tA}f)(x,y) \equiv f(x + t, y)$, and $(Bf)(x,y) \equiv h(y)f(x,y)$, where $h \in W^{m,\infty}(\mathbf{R}^2)$. Then $G \equiv BA$ is given by

$$(Gf)(x,y) \equiv h(y)\frac{\partial f}{\partial x}(x,y).$$

Proposition.

  (a) *There exists bounded injective C with dense range, such that G generates an entire C-regularized group.*

  (b) *If h is real-valued, then G generates a bounded $(1 + A)^{-(m+1)}$-regularized semigroup.*

  (c) *If h is real-valued and $\frac{1}{h} \in W^{m,\infty}(\mathbf{R}^2)$, then G generates an exponentially bounded $(m + 1)$-times integrated semigroup.*

Proof: (a) is clear from Theorem 19.6 (a). (b) follows from Theorem 19.6 (b) and the fact that $(e^{itB}f)(x,y) = e^{ith(y)}f(x,y)$, so that $\|e^{itB}\|$ is $O\big(\sum_{k=0}^m \|(d/dy)^k e^{ith(y)}\|\big)$, which is $O(t^m)$, since $h \in W^{m,\infty}(\mathbf{R}^2)$, and $|e^{ith(y)}| = 1$, for all t, y. (c) is clear from (b) and Theorem 19.6(c). ∎

Before proving our theorems, let us give an example of just how bad a bounded, additive perturbation can be.

Example 19.11. This is an example of operators A, B, such that the spectrum of A is contained in the imaginary axis, the domain of A is dense, A generates a bounded $(1 + A)^{-1}$-regularized group and an exponentially bounded once-integrated group, and B is bounded, but $(A + B)$ does not generate a C-regularized semigroup, for any C (hence does not generate an n-times integrated semigroup, for any n; see Theorem 18.3). Let X be the bumpy translation space of Example 4.17,

$$X \equiv C_0(\mathbf{R}) \bigcap C_0^1([0, \infty)),$$

119

where

$$C_0^1([0,\infty)) \equiv \{f \in C^1([0,\infty)) \mid \lim_{x\to\infty} f(x) = 0 = \lim_{x\to\infty} f'(x)\},$$

with

$$\|f\| \equiv \|f\|_{C_0(\mathbf{R})} + \|f'\|_{C_0([0,\infty))}.$$

As in Example 4.17, let $A \equiv -\frac{d}{dx}$, the generator of right-translation on $\mathbf{R}$, $Bf \equiv qf$, for some $q \in X$ such that q is not differentiable at any point in $(-\infty, 0)$.

In Example 4.18, we showed that A generates a bounded $(1 + A)^{-1}$-regularized group and an exponentially bounded once-integrated group.

Let us write $u(t, f)$ for a mild solution of $u' = (A + B)u, u(0, f) = f$. These have the form

$$[u(t,f)](x) = \exp\left[\int_0^t q(x - s)\, ds\right] f(x - t) \quad (x \in \mathbf{R});$$

these solutions are reversible, that is, t can be any real number.

Suppose that $A + B$ generates a C-regularized semigroup. Since C commutes with $A + B$ (see Theorem 3.4), $C(u(t,f)) = u(t, Cf)$, for any f in the solution space of $A + B$, t real. We also showed, in Example 4.18, that the solution space for $A + B$ included only functions supported in $[0, \infty)$. Thus, the image of C includes only functions supported in $[0, \infty)$. We claim that this implies that the image of C is trivial.

Suppose $f \in Im(C)$. For $x < 0, t \in \mathbf{R}$,

$$0 = [C(u(t,f))](x) = [u(t, Cf)](x)$$
$$= \exp\left[\int_0^t q(x-s)\, ds\right] Cf(x-t),$$

so that

$$Cf(x - t) = 0, \ \forall x < 0, t \in \mathbf{R}.$$

This clearly implies that $Cf = 0$. Since C is injective, this implies that $f = 0$, that is, the image of C is trivial.

Since C must be injective, this is impossible, thus we conclude that $A + B$ cannot generate a C-regularized semigroup, for any C.

Proof of Theorem 19.2: Let Z be the Hille-Yosida space for A, $Z_0(A)$ (see Definition 5.1). By Proposition 5.15, W, of (4), is $[C_1^{-1}C_2(Z)]$. Since C_2 commutes with $W(t)$, it commutes with $e^{tA|z}$, thus $A|_W$ also generates a strongly continuous contraction semigroup. Since $W \hookrightarrow X$, $B|_W$

is closed in W. Thus, by hypothesis (4) and the Closed Graph Theorem, $B|_W \in B(W)$. This implies that $(A|_W + B|_W)$ generates a strongly continuous semigroup $\{T(t)\}_{t\geq 0}$, on W. This is saying that W is contained in $Z_\omega(A+B)$, for some $\omega \in \mathbf{R}$, so that, by Theorem 5.16, since $Im(C_2) \subseteq W$, the conclusion follows. ∎

Proof of Theorem 19.4: For $s \in \rho_{C_1}(A)$, define

$$R(s) \equiv \sum_{k=0}^{N-1} \left[(s-A)^{-1}C_1 BC_2\right]^k \left((s-A)^{-1}C_1\right)(C_1 C_2)^{N-k}.$$

By (2) and (4), $R(s)x \in \mathcal{D}(A+B)$, for all $x \in X$. Let $C \equiv C_1(C_1 C_2)^N$. We will show that $(s-(A+B))^{-1}C$ exists and equals $R(s)$. For $x \in \mathcal{D}(A+B)$, the calculation follows:

$$R(s)(s-(A+B))x$$

$$= C_1 \sum_{k=0}^{N-1} \left[(s-A)^{-1}C_1 BC_2\right]^k (C_1 C_2)^{N-k}x$$

$$- C_1 \sum_{k=0}^{N-1} \left[(s-A)^{-1}C_1 BC_2\right]^k \left((s-A)^{-1}C_1 BC_2\right)(C_1 C_2)^{N-k-1}x$$

$$= C_1 \left[(C_1 C_2)^N x - \left[(s-A)^{-1}C_1 BC_2\right]^N x\right]$$

$$= Cx.$$

Thus, $(s-(A+B))$ is injective, and, for all $x \in X$,

$$R(s)(s-(A+B))R(s)x = CR(s)x = R(s)Cx,$$

so that, since $R(s)$ is injective,

$$(s-(A+B))R(s)x = Cx.$$

This implies that $\rho_{C_1}(A) \subseteq \rho_C(A+B)$, with, for $s \in \rho_{C_1}(A)$,

$$R(s) = (s-(A+B))^{-1}C,$$

$$\|(s-(A+B))^{-1}C\| \leq \sum_{k=0}^{N} \|(s-A)^{-1}C_1\|^{k+1}\|BC_2\|^k\|C_1 C_2\|^{N-k}.$$

Theorem 17.7 and Proposition 17.2 now conclude the proof. ∎

Our technique in proving Theorem 19.6 is to use the Fourier inversion formula to construct the C-regularized semigroup generated by BA, which may be thought of as

$$e^{sBA}C \equiv \int_{\mathbf{R}} e^{itA} \mathcal{F}(f_s)(t)\, dt,$$

where $f_s(r) \equiv g(r)e^{srB}$,

$$C = \int e^{itA} \mathcal{F}(g)(t)\, dt,$$

and g is a suitable function; the choice of g will depend on the behaviour of $\{e^{sB}\}_{s\geq 0}$. (In this formula, (iA) generates a bounded strongly continuous group.)

To prove Theorem 19.6 (c), we will need the following.

Lemma 19.12. *Suppose A and B are as in Theorem 19.6 (c). Then $(0,\infty) \subseteq \rho(BA)$.*

Proof: Since $\{\|e^{itB}\|\}_{t\in\mathbf{R}}$ is $O(t^n)$, for t large, the spectrum of B is real (see [Dav1]). Thus, the hypotheses on B imply that $\operatorname{sp}(B) \subseteq \mathbf{R} - \{0\}$. For $s > 0$, define the bounded operator

$$R(s) \equiv \int_{\Gamma} (s - wA)^{-1}(w - B)^{-1} \frac{dw}{2\pi i},$$

where Γ is a cycle that bounds $\operatorname{sp}(B)$ and does not intersect the imaginary axis.

Note that $(s - wA)^{-1} = 1/w\big((s\overline{w})/|w|^2 - A\big)^{-1}$ exists and $w \mapsto (s - wA)^{-1}$ is a holomorphic map, from $(\mathbf{C} - i\mathbf{R})$ into $B(X)$, since $\operatorname{sp}(A) \subseteq i\mathbf{R}$. Since Γ is bounded, and $(s - wA)^{-1}x \in \mathcal{D}(A)$, for all $w \in \Gamma$, $R(s)x \in \mathcal{D}(A)$, for all $x \in X$. Hence, we may calculate as follows, for $x \in X$:

$$2\pi i(s - BA)R(s)x = \int_{\Gamma} (s - BA)(s - wA)^{-1}(w - B)^{-1}x\, dw$$

$$= \int_{\Gamma} \big(s - wA + (w - B)A\big)(s - wA)^{-1}(w - B)^{-1}x\, dw$$

$$= \int_{\Gamma} (w - B)^{-1}x\, dw + \int_{\Gamma} A(s - wA)^{-1}x\, dw$$

$$= 2\pi i x,$$

by Cauchy's theorem, and the usual Gelfand functional calculus for bounded operators.

Thus $(s - BA)R(s)x = x$, for all $x \in X$. It is clear that, for $x \in \mathcal{D}(A)$, $R(s)(s - BA)x = (s - BA)R(s)x = x$. Thus $s \in \rho(BA)$, with $(s - BA)^{-1} = R(s)$.

Proof of Theorem 19.6: It is sufficient to assume that iA generates a bounded strongly continuous group, and, in (b) and (c), $\{\|e^{tB}\|\}_{t \in \mathbf{R}}$ is $O(t^n)$.

Let

$$\mathcal{A} \equiv \{\text{strongly continuous } f : \mathbf{R} \to B(X) \,|\, \forall x \in X, \text{ the maps}$$

$$r \mapsto f(r)x \text{ and } r \mapsto f'(r)x \text{ are in } L^2(\mathbf{R}, X)\}.$$

For $f \in \mathcal{A}$, define the operator $f(A)$ by, for x in X,

$$f(A)x \equiv \int_{\mathbf{R}} \lim_{N \to \infty} \int_{-N}^{N} e^{-irt} e^{itA} f(r) x \frac{dr}{2\pi}\, dt, \qquad (19.13)$$

where the limit is taken in $L^2(\mathbf{R}, X)$.

When f is scalar-valued, this functional calculus appears in [Dav1, Chapter 8]. The same arguments that appear there may be used to show that

$$f(A)g(A) = (fg)(A), \quad \forall f, g \in \mathcal{A}; \qquad (19.14)$$

$$\exists M < \infty \text{ such that } \|f(A)x\| \le M(\|f(r)x\|_2 + \|f'(r)x\|_2), \quad \forall x \in X; \qquad (19.15)$$

$$(i + A)^{-1} = g(A), \quad \text{where } g(r) \equiv (i + r)^{-1}. \qquad (19.16)$$

For (a), define

$$W_1(z) \equiv g_z(A), \quad \text{where } g_z(r) \equiv e^{zrB} e^{-r^2} \ (z \in \mathbf{C}, r \in \mathbf{R}).$$

For (b), define

$$W_2(s) \equiv f_s(A), \quad \text{where } f_s(r) \equiv \frac{e^{srB}}{(i + r)^{n+1}}.$$

Assertion (19.15) implies that $\{W_2(s)\}_{s \ge 0}$ is an exponentially bounded strongly continuous family of bounded operators and $\{W_1(z)\}_{z \in \mathbf{C}}$ is an entire family of bounded operators. Assertions (19.14) and (19.16) imply that $\{W_2(s)\}_{s \ge 0}$ is an $(i + A)^{-(n+1)}$-regularized semigroup, while (19.14)

123

implies that $\{W_1(z)\}_{z\in\mathbb{C}}$ is an entire $W_1(0)$-group; it was shown in Lemma 8.8 that $C \equiv W_1(0) = e^{-A^2}$, is injective and has dense range.

For (a) and (b), all that remains is to show that $\{W_1(z)\}_{z\in\mathbb{C}}$ and $\{W_2(s)\}_{s\geq 0}$ are generated by BA. (19.14) and (19.16) allow us to make the following calculations, for $x \in X$:

$$\frac{d}{ds}W_2(s)(i + A)^{-1}x = \int_{\mathbb{R}} \lim_{N\to\infty} \int_{-N}^{N} e^{-irt}e^{itA}\frac{d}{ds}\left(\frac{e^{srB}x}{(i+r)^{n+2}}\right)\frac{dr}{2\pi}\, dt$$

$$= \int_{\mathbb{R}} \lim_{N\to\infty} \int_{-N}^{N} e^{-irt}e^{itA}Be^{srB}x\left(\frac{i+r-i}{(i+r)^{n+2}}\right)\frac{dr}{2\pi}\, dt$$

$$= B[W_2(s)x - iW_2(s)(i + A)^{-1}x]$$

$$= BAW_2(s)(i + A)^{-1}x,$$

$$\tag{19.17}$$

$$\frac{d}{dz}W_1(z)(i + A)^{-1}x = BAW_1(z)(i + A)^{-1}x, \tag{19.18}$$

exactly as with (19.17).

Assertion (19.17) implies that an extension of BA generates $\{W_2(s)\}_{s\geq 0}$ and (19.18) implies that an extension of BA generates $\{W_1(z)\}_{z\in\mathbb{C}}$. (a) and (b) now follow from Proposition 3.9, with $G = (i + A)^{-1}$.

Since $(i + A)(i + BA)^{-1}$ is bounded and injective, (b) implies that BA generates an exponentially bounded $(i+BA)^{-(n+1)}$-regularized semigroup. By Theorem 18.3, BA generates an exponentially bounded $(n+1)$-times integrated semigroup, proving (c). ∎

XX. TYPE OF AN OPERATOR

In considering the spectrum of an operator, A, the goal is to think of A as a complex number, or a set of complex numbers. For example, if one wants A to generate a bounded, strongly continuous semigroup, which we think of as e^{tA}, we want the spectrum of A to be contained in the left half-plane $Re(z) \leq 0$, since this is the set of all z for which $t \mapsto e^{tz}$, from $[0, \infty)$ into the complex plane, is bounded. Throughout this book, we have constructed $g(A)$-regularized semigroups $W(t)$, generated by A, by thinking of $W(t)$ as $e^{tA}g(A)$, with g chosen so that the map $t \mapsto e^{tz}g(z)$ is bounded, for all z in the spectrum of A.

But it is well known that specifying the *location* of the spectrum is not enough. For generation theorems, it is also necessary to have conditions on the rate of growth of the resolvents. For example, the Hille-Yosida theorem states that a densely defined operator A generates a bounded strongly continuous semigroup if and only if the spectrum is contained in the closed left half-plane and

$$\{\|Re(z)(z - A)^{-1}\| \mid Re(z) < 0\}$$

is bounded.

One very successful approach is to specify both the spectrum and the *numerical range* (see any of the references on strongly continuous semigroups). The Lumer-Phillips theorem states that a densely defined operator generates a strongly continuous semigroup of contractions if and only if the spectrum and numerical range are contained in the closed left half-plane.

There are two disadvantages to using the numerical range. First, it does not, in general, satisfy the sort of *mapping theorems* that the spectrum does; that is, it is not true in general that $n.r.(f(A)) = f(n.r.(A))$, even when f is chosen to be a polynomial. Consider, on $\mathbf{C}^2$,

$$A \equiv \begin{bmatrix} 4 + i & 4i \\ 4i & 16 + 4i \end{bmatrix}.$$

Then a calculation shows that the numerical range of A is contained in $S_{\frac{\pi}{4}} \equiv \{re^{i\phi} \mid r > 0, |\phi| < \frac{\pi}{4}\}$, but the numerical range of A^2 is not contained in $S_{\frac{\pi}{2}}$. Thus, if $f(z) \equiv z^2$, then $n.r.(f(A))$ is not contained in $f(n.r.(A))$. Second, there does not appear to be a reasonable analogue of the numerical range for the more general classes of semigroups that we are considering, such as regularized semigroups.

We would like, associated with any operator, a set of complex numbers, with the following properties:

(1) it satisfies mapping theorems;

(2) its location tells us something about generation of regularized semi-groups.

Throughout this chapter, V and O will be open subsets of the complex plane whose complement contains a half-line and whose boundaries, $\partial V, \partial O$, are positively oriented countable systems of piecewise smooth, mutually nonintersecting (possibly unbounded) arcs.

We will write RHP for the open right half-plane $\{z \in \mathbf{C} \mid Re(z) > 0\}$, LHP for the open left half-plane, $S_\phi \equiv \{re^{i\psi} \mid r > 0, |\psi| < \phi\}(0 < \phi \leq \pi)$, $H_\epsilon \equiv \{z \in \mathbf{C} \mid |Im(z)| < \epsilon\}$.

Definition 20.1. Suppose $\alpha \geq -1$. We will say that the operator B is of α-type V if $sp(B) \subseteq V$ and $\|(w - B)^{-1}\|$ is $O((1 + |w|)^\alpha)$, for $w \notin V$.

Examples 20.2. (1) If $B \in B(X)$, then B is of (-1)-type V, whenever V is an open set containing $sp(B)$.

(2) Let $(Bf)(z) \equiv zf(z)$, on $L^2([1, \infty))$. Then, for all $\theta > 0$, B is of (-1)-type S_θ. for all $\epsilon > 0$, B is of 0-type H_ϵ, but is not of (-1)-type H_ϵ.

This illustrates the fact, that, although many operators are of (-1)-type V, for *some* V, many natural choices of V require that we consider operators of α-type V, for $\alpha > -1$.

Another example of this is $B \equiv -\Delta$, the Laplacian, on $L^p(\mathbf{R}^n), 1 \leq p \leq \infty$, with $\mathcal{D}(B) = \{f \in L^p(\mathbf{R}^n) \mid \Delta f \in L^p(\mathbf{R}^n)\}$, where Δ is taken in the sense of distribution. In particular, if $1 < p < \infty$, then $\mathcal{D}(B) = W^{2,p}(\mathbf{R}^n)$ (the Sobolev space)(see [H-V]). For all $\theta > 0$, B is of (-1)-type $(S_\theta - 1)$, while for all $\epsilon > 0$, there exists $\alpha > -1$, depending on both p and n, such that B is of α-type H_ϵ, but B is not of (-1)-type H_ϵ (this may be seen by using the functional calculus for commuting generators of bounded strongly continuous groups; see Chapter XII).

Consider also an operator, A, that generates an exponentially decaying strongly continuous semigroup, $\{e^{tA}\}_{t \geq 0}$. The operator A is of 0-type LHP and is of (-1)-type LHP if and only if $\{e^{tA}\}$ extends to a bounded holomorphic strongly continuous semigroup.

The following example provides a prototype operator of α-type RHP.

(3) For any $\alpha \geq -1, k > 0$, let $O_{\alpha,k} \equiv \{x + iy \mid x > k(1 + |y|)^{-\alpha}\}, V \equiv RHP$, $(B_{\alpha,k}f)(z) \equiv zf(z)$, on $L^2(\overline{O_{\alpha,k}})$. Then, for all $k > 0, B_{\alpha,k}$ is of α-type V and is not of β-type V, for any $\beta > \alpha$.

(4) Let Γ be the unit circle in the complex plane and let $B \equiv \frac{d}{d\theta}$, the generator of rotation on $L^p(\Gamma), 1 \leq p \leq \infty$, V defined to be a union of

open balls of radius less than one, centered at the integers. Then iB is of 0-type V.

(5) If $\mathcal{D}(B)$ is dense, then it is well known that B is of (-1)-type S_ϕ, for all $\phi > \theta$, if and only if $(-B)$ generates an exponentially decaying holomorphic strongly continuous semigroup of angle $(\frac{\pi}{2} - \theta)(0 \le \theta < \frac{\pi}{2})$.

(6) More generally, if $(m+2) \in \mathbf{N}$, $\lambda \in \rho(B)$ and $\mathcal{D}(B)$ is dense, then B is of m-type S_ϕ, for all $\phi > \theta$, if and only if $(-B)$ generates an exponentially decaying holomorphic $(\lambda - B)^{-(m+1)}$-regularized semigroup of angle $(\frac{\pi}{2} - \theta)$ (see the next chapter).

Hence, being of m-type V is analogous to having the numerical range contained in V; the desired property here is generating a $(\lambda - B)^{-(m+1)}$-regularized semigroup rather than a strongly continuous contraction semigroup.

In terms of the abstract Cauchy problem, being of m-type $-S_\theta$ $(\theta < \frac{\pi}{2})$ corresponds to the abstract Cauchy problem having a unique exponentially decaying solution, for all $x \in \mathcal{D}(B^{m+1})$, while having the numerical range contained in $-S_\theta$ corresponds to the abstract Cauchy problem having a unique nonincreasing solution, for all $x \in \mathcal{D}(B)$.

(7) In many places ([B-C], [Bo-dL3], [Do-V1, 2], [Duo1, 2], [M], [M-Y], [Pr-Soh], [Ri1, 2], [V] and [Y1, 2], for example), operators of *type* ω are considered. These are densely defined operators, A, such that $sp(A) \subseteq \overline{S_\omega}$ and for all $\psi > \omega$, there exists $M_\psi < \infty$ such that $\|w(w - A)^{-1}\| \le M_\psi$, for all $w \notin S_\psi$. Note that a densely defined operator with 0 in its resolvent set is of type ω if and only if it is of (-1)-type S_ϕ, for all $\phi > \omega$.

Even on a Hilbert space, when $f \in H^\infty(S_\phi)$ and B is of type ϕ, $f(B)$ may not be bounded (see [M-Y] and [B-C]).

(8) In [Be1, 2] operators of n-type V_a, where $n \in \mathbf{N}$ is arbitrary, $0 < a < 1$ and $V_a \equiv \{x + iy|\, x < |y|^a\}$ are considered.

A simple example of such an operator (from [Be1]) is

$$B \equiv \begin{bmatrix} G & G \\ 1 & G \end{bmatrix},$$

on $X \times X$, $\mathcal{D}(B) \equiv \mathcal{D}(G) \times \mathcal{D}(G)$, where G generates a strongly continuous semigroup.

XXI. HOLOMORPHIC C-EXISTENCE FAMILIES

In this chapter, we define exponentially bounded holomorphic existence families (Definition 21.5) so as to generalize holomorphic strongly continuous semigroups, a class of semigroups that has found wide applicability.

In keeping with our policy, in this book, of not presenting proofs that are obvious modifications of proofs for strongly continuous semigroups, we have included no more of the proofs than the basic construction of the desired family of operators.

Our main results may be summarized as follows. For an operator, A, to generate an exponentially bounded holomorphic k-times integrated semigroup, it is sufficient that it be of $(k-1-\epsilon)$-type V, for an appropriate sector V, for some $\epsilon > 0$ (see Definition 21.1). If A is densely defined, it is necessary and sufficient that it be of $(k-1)$-type V. In order that there exist an exponentially bounded holomorphic C-existence family for A, it is sufficient that $\|A(w-A)^{-1}C\|$ be $O(|w|^{-\epsilon})$, for some positive ϵ (Theorems 21.8 and 21.9). If A is densely defined and $CA \subseteq AC$ (this is automatically true when A generates a C-regularized semigroup), it is necessary and sufficient that $\|(A(w-A)^{-1}C\|$ be bounded (Theorems 21.15 and 21.16).

Definition 21.1. $S_\Theta \equiv \{re^{i\phi}|r > 0, |\phi| < \Theta\}$, $V_\Theta \equiv \{re^{i\phi}|r > 0, |\phi| \leq \Theta\}$.

Definition 21.2. Suppose $\frac{\pi}{2} \geq \Theta > 0$. The n-times integrated semigroup $\{S(t)\}_{t\geq 0}$ is a *holomorphic n-times integrated semigroup of angle* Θ if it extends to a family of bounded operators $\{S(z)\}_{z\in S_\Theta}$ satisfying:

(1) the map $z \mapsto S(z)$, from S_Θ into $B(X)$, is holomorphic;

(2) $\{\left(\frac{d}{dz}\right)^n S(z)\}_{z\in S_\Theta}$ is a semigroup; and

(3) for all $\psi < \Theta$, $\{S(z)\}$ is strongly continuous on $\overline{S_\psi}$.

Definition 21.3. Suppose $\frac{\pi}{2} \geq \Theta > 0$. The C-regularized semigroup $\{W(t)\}_{t\geq 0}$ is a *holomorphic C-regularized semigroup of angle* Θ if it extends to a family of bounded operators $\{W(z)\}_{z\in S_\Theta}$ satisfying:

(1) the map $z \mapsto W(z)$, from S_Θ into $B(X)$, is holomorphic;

(2) $W(z)W(w) = CW(z+w)$, for all $z, w \in S_\Theta$; and

(3) for all $\psi < \Theta$, $\{W(z)\}$ is strongly continuous on $\overline{S_\psi}$.

Definition 21.4. The family of operators in Definition 21.2 (21.3) is *exponentially bounded* if, for all $\psi < \Theta$, there exist finite M_ψ, w_ψ, such that $\|S(z)\| (\|W(z)\|) \leq M_\psi e^{w_\psi |z|}$, for all $z \in S_\psi$.

128

When $w_\psi = 0$, for all $\psi < \Theta$, then the family of operators is *bounded*. Note that M_ψ may get arbitrarily large as ψ gets close to Θ.

Definition 21.5. Suppose $\frac{\pi}{2} \geq \Theta > 0$. The exponentially bounded mild C-existence family $\{W(t)\}_{t \geq 0}$ is an *exponentially bounded holomorphic C-existence family of angle Θ for A* if it extends to a family of bounded operators $\{W(z)\}_{z \in S_\Theta}$ satisfying:

(1) the map $z \mapsto W(z)$, from S_Θ into $B(X)$, is holomorphic;

(2) whenever $|\phi| < \Theta$, $\{W(te^{i\phi})\}_{t \geq 0}$ is an exponentially bounded mild C-existence family for $e^{i\phi} A$; and

(3) for all $\psi < \Theta$, $\{W(z)\}$ is strongly continuous on $\overline{S_\psi}$.

If $\|W(z)\|$ is bounded on $\overline{S_\psi}$, for all $\psi < \Theta$, then $\{W(z)\}_{z \in S_\Theta}$ is a *bounded holomorphic mild C-existence family*.

Theorem 21.6. *Suppose A is closed, $\frac{\pi}{2} \geq \Theta > 0$, $S_{(\Theta + \frac{\pi}{2})} \subseteq \rho_C(A)$, C is injective and commutes with $(w - A)^{-1}C$, for all $w \in \rho_C(A)$ and $\{W(z)\}_{z \in S_\Theta} \subseteq B(X)$. Then the following are equivalent.*

(a) $\{W(z)\}_{z \in S_\Theta}$ *is an exponentially bounded holomorphic C-regularized semigroup of angle Θ generated by an extension of A.*

(b) $\{W(z)\}_{z \in S_\Theta}$ *is an exponentially bounded holomorphic mild C-existence family of angle Θ for A.*

Theorem 21.7. *Suppose there exists real r such that $[r, \infty) \subseteq \rho(A)$, and $\frac{\pi}{2} \geq \Theta > 0$. Then the following are equivalent.*

(a) *The operator A generates an exponentially bounded holomorphic n-times integrated semigroup $\{S(z)\}_{z \in S_\Theta}$, of angle Θ.*

(b) *The operator A generates an exponentially bounded holomorphic $(A - r)^{-n}$- semigroup, $\{W(z)\}_{z \in S_\Theta}$, of angle Θ.*

(c) *There exists a holomorphic semigroup $\{T(z)\}_{z \in S_\Theta}$ satisfying the following.*

(1) If $z \in S_\Theta$ and $x \in \mathcal{D}(A)$, then $T(z)x \in \mathcal{D}(A)$, with $\frac{d}{dz} T(z)x = AT(z)x = T(z)Ax$.

(2) If $\psi < \Theta$ and $x \in \mathcal{D}(A^n)$, then $T(z)x$ converges to x, as $z \to 0$ in $\overline{S_\Theta}$.

(3) For all $\psi < \Theta$, there exists finite M_ψ, w_ψ, such that $\|T(z)x\| \leq M_\psi e^{w_\psi |z|} \|(A - r)^n x\|$, for all $x \in \mathcal{D}(A^n)$, $z \in S_\Theta$.

We then have $\left(\frac{d}{dz}\right)^n S(z) = T(z)$, $W(z) = (A - r)^{-n} T(z)$, for $z \in S_\Theta$.

Theorem 21.8. *Suppose A is closed and there exists $\pi > \psi > \frac{\pi}{2}$ such that $V_\psi \subseteq \rho_C(A)$, and $w \mapsto (w - A)^{-1}$, from V_ψ into $B(X)$, is holomorphic.*

129

Then there exists a bounded holomorphic mild C-existence family of angle $(\psi - \frac{\pi}{2})$ for A if either

(a) $\{x \in \mathcal{D}(AC) \mid ACx \in Im(C)\}$ is dense and $\|A(w - A)^{-1}C\|$ is bounded in V_ψ, or

(b) there exists $\epsilon > 0$ such that $\|A(w - A)^{-1}C\|$ is bounded and $O\left(|w|^{-\epsilon}\right)$ in V_ψ.

Theorem 21.9. *Suppose A is closed and there exist $\pi > \psi > \frac{\pi}{2}, k > 0$ such that $(k + V_\psi) \subseteq \rho_C(A)$, and $w \mapsto (w - A)^{-1}C$, from $(k + V_\psi)$ into $B(X)$, is holomorphic. Then there exists an exponentially bounded holomorphic mild C-existence family of angle $(\psi - \frac{\pi}{2})$ for A if either*

(a) $\{x \in \mathcal{D}(AC) \mid ACx \in Im(C)\}$ *is dense and* $\|A(w - A)^{-1}C\|$ *is bounded in* $(k + V_\psi)$, *or*

(b) *there exists $\epsilon > 0$ such that* $\|A(w - A)^{-1}C\|$ *is* $O\left(|w|^{-\epsilon}\right)$ *in* $(k + V_\psi)$.

Corollary 21.10. *Suppose A is closed and there exists $\pi > \psi > \frac{\pi}{2}$ such that $V_\psi \subseteq \rho_C(A)$, the map $w \mapsto (w - A)^{-1}C$ is holomorphic and C is injective and commutes with $(w - A)^{-1}C$, for all $w \in V_\psi$. Then an extension of A generates a bounded holomorphic C-regularized semigroup of angle $(\psi - \frac{\pi}{2})$ that leaves $\mathcal{D}(A)$ invariant if either*

(a) $\mathcal{D}(A)$ *is dense and* $\|A(w - A)^{-1}C\|$ *is bounded in* V_ψ, *or*

(b) *there exists $\epsilon > 0$ such that* $\|A(w - A)^{-1}C\|$ *is bounded and* $O\left(|w|^{-\epsilon}\right)$ *in* V_ψ.

Corollary 21.11. *Suppose A is closed and there exist $\pi > \psi > \frac{\pi}{2}, k > 0$, such that $(k + V_\psi) \subseteq \rho_C(A)$, the map $w \mapsto (w - A)^{-1}C$ is holomorphic and C is injective and commutes with $(w - A)^{-1}C$, for all $w \in V_\psi$. Then an extension of A generates an exponentially bounded holomorphic C-regularized semigroup of angle $(\psi - \frac{\pi}{2})$ that leaves $\mathcal{D}(A)$ invariant if either*

(a) $\mathcal{D}(A)$ *is dense and* $\|A(w - A)^{-1}C\|$ *is bounded in* $(k + V_\psi)$, *or*

(b) *there exists $\epsilon > 0$ such that* $\|A(w - A)^{-1}C\|$ *is* $O\left(|w|^{-\epsilon}\right)$ *in* $(k + V_\psi)$.

Remark 21.12. In the preceding results, in order that the map $w \mapsto (w - A)^{-1}C$, from V_ψ into $B(X)$, be holomorphic, it is sufficient to have A closed and $Im(C) \subseteq Im\left((w - A)^3\right)$, for $w \in V_\psi$, with $\|(w - A)^{-1}(r - A)^{-1}(s - A)^{-1}\|$ locally bounded. This may be shown with the identity

$$(r - A)^{-1}C - (s - A)^{-1}C = (s - r)(r - A)^{-1}(s - A)^{-1}C.$$

Note that 21.13(b) is similar to Theorem 17.7.

Theorem 21.13. *Suppose there exist* $\pi > \psi > \frac{\pi}{2}, k > 0$, *such that* $(k + \overline{S_\psi}) \subset \rho(A)$. *Then A generates an exponentially bounded holomorphic n-times integrated semigroup of angle $(\psi - \frac{\pi}{2})$ if either*

(a) *$\mathcal{D}(A)$ is dense and A is of $(n-1)$-type $(k + \overline{S_\psi})$, or*

(b) *there exists $\epsilon > 0$ such that A is of $(n - 1 - \epsilon)$-type $(k + \overline{S_\psi})$.*

Proof of Theorem 21.8: For $r > 0$, let

$$\Gamma_r \equiv \{se^{\pm i\psi} \mid s \geq r\} \cup \{re^{i\theta} \mid -\psi \leq \theta \leq \psi\},$$

oriented counterclockwise. Define, for $z \in S_{(\psi - \frac{\pi}{2})}$,

$$W(z) \equiv \int_{\Gamma_r} e^{zw}(w - A)^{-1} C \, \frac{dw}{2\pi i}.$$

By Cauchy's theorem, this definition is independent of $r > 0$.

The construction of $T(z)$ relies on the following.

Lemma 21.14. *Suppose k and r are nonnegative, $\psi > \frac{\pi}{2}, (k + \overline{S_\psi}) \subseteq \rho(A), |\arg(z)| < (\psi - \frac{\pi}{2})$, and n is a nonnegative integer. Let $\Gamma_{r,k} \equiv k + \Gamma_r$, where Γ_r is defined in the proof of Theorem 21.8. Then, for $x \in \mathcal{D}(A^n)$,*

$$\int_{\Gamma_{r,k}} e^{zw}(w - A)^{-1}x \, dw = \int_{\Gamma_{r,k}} e^{zw}(w - A)^{-1}A^n x \, \frac{dw}{w^n} + 2\pi i \sum_{j=0}^{n-1} \frac{z^j}{j!} A^j x.$$

Proof of Theorem 21.13: With $\Gamma_{r,k}$ as in Lemma 21.14, $x \in X$, define

$$T(z)x \equiv \int_{\Gamma_{r,k}} e^{zw}(w - A)^{-1}x \, \frac{dw}{2\pi i},$$

for $|\arg(z)| < (\psi - \frac{\pi}{2}), r > 0$.

Theorem 21.15. *Suppose A is closed, $\{x \in \mathcal{D}(AC) \mid ACx \in Im(C)\}$ is dense, $\frac{\pi}{2} \geq \Theta > 0$ and $S_{(\frac{\pi}{2} + \Theta)} \subseteq \rho_C(A)$. Then the following are equivalent.*

(a) *There exists a bounded holomorphic mild C-existence family of angle Θ for A.*

(b) *For all $\psi < (\frac{\pi}{2} + \Theta), \|A(w - A)^{-1}C\|$ is bounded in S_ψ.*

131

Theorem 21.16. *Suppose A is closed, $\{x \in \mathcal{D}(AC) \mid ACx \in Im(C)\}$ is dense, $\frac{\pi}{2} \geq \Theta > 0$ and for all $\psi < (\frac{\pi}{2} + \Theta)$, there exists $k_\psi \in \mathbf{R}$ such that $(k_\psi + S_\psi) \subseteq \rho_C(A)$. Then the following are equivalent.*

  (a) *There exists an exponentially bounded holomorphic mild C-existence family of angle Θ for A.*

  (b) *For all $\psi < (\frac{\pi}{2} + \Theta)$, $\|A(w - A)^{-1}C\|$ is bounded in $(k_\psi + S_\psi)$.*

When A generates a bounded holomorphic C-regularized semigroup of angle Θ, then $S_{\frac{\pi}{2}+\Theta} \subseteq \rho_C(A)$; this is a consequence of Proposition 17.2. When there exists a bounded holomorphic mild C-existence family of angle Θ for A, then it may be shown that $Im(C) \subseteq Im(w - A)$, for all $w \in S_{\frac{\pi}{2}+\Theta}$.

Theorem 21.17. *Suppose $\mathcal{D}(A)$ is dense. Then the following are equivalent.*

  (a) *The operator A generates an exponentially bounded n-times integrated semigroup of angle Θ.*

  (b) *For all $\psi < (\frac{\pi}{2} + \Theta)$, there exists $k_\psi > 0$ such that $(k_\psi + S_\psi) \subseteq \rho(A)$ and $\|(A - r)^{-n}A(w - A)^{-1}\|$ is bounded in $(k_\psi + S_\psi)$, for some $r \in \rho(A)$.*

  (c) *For all $\psi < (\frac{\pi}{2} + \Theta)$, there exists $k_\psi > 0$ such that A is of $(n-1)$-type $(k_\psi + S_\psi)$.*

As in Examples 16.9 and 17.9, we have the following, where we use the same terminology as in 17.9.

Example 21.18. *Suppose $n, m \in \mathbf{N} \cup 0, s \in \rho(G_1) \cap \rho(G_2)$,*

$$A \equiv \begin{bmatrix} G_1 & B \\ 0 & G_2 \end{bmatrix}, \quad \mathcal{D}(A) \equiv \mathcal{D}(G_1) \times [\mathcal{D}(B) \cap \mathcal{D}(G_2)],$$

where

  (1) *G_i generates a strongly continuous holomorphic semigroup, for $i = 1, 2$;*

  (2) *$(s - G_1)^{-m}B$ is bounded;*

  (3) *$\mathcal{D}(G_2^n) \subseteq \mathcal{D}(B)$; and*

  (4) *B is closed.*

Then A is closable and there exists an exponentially bounded holomorphic mild C-existence family for $\overline{A}$, where

$$C \equiv \begin{bmatrix} I & 0 \\ 0 & (s - G_2)^{-n} \end{bmatrix}.$$

XXII. UNBOUNDED HOLOMORPHIC FUNCTIONAL CALCULUS FOR OPERATORS WITH POLYNOMIALLY BOUNDED RESOLVENTS

In studying linear operators, it is desirable to gain as much information as possible by looking at the spectrum. Perhaps the most well-known functional calculus for arbitrary bounded operators on a Banach space is the Riesz-Dunford functional calculus,

$$f(A) \equiv \int_\Gamma f(w)(w - A)^{-1} \frac{dw}{2\pi i}, \qquad (22.1)$$

where f is holomorphic in an open neighborhood containing the spectrum of A and Γ surrounds the spectrum of A.

Of more interest in applications are unbounded operators, such as differential operators. Two problems immediately arise when one tries to extend (22.1) to unbounded operators. The spectrum of A may be unbounded and functions holomorphic on the spectrum of A may be unbounded.

Since $\int_\Gamma \|(w - A)^{-1}\| \, d|w|$ is no longer finite, it is not surprising that, even for bounded holomorphic f, $f(A)$ may not be bounded, even when $\|(w - A)^{-1}\|$ is $O(\frac{1}{1+|w|})$, as with a bounded operator. For example, let D be the open unit disc in the complex plane, let $-iA$ be $\frac{d}{d\theta}$, on $L^1(\partial D)$, the generator of the rotation group on ∂D and let $f \equiv 1_{[0,\infty)}$. Then it is well known that $f(A)$, the projection of $L^1(\partial D)$ onto $H^1(D)$, is unbounded. Even on a Hilbert space, $f(A)$ may be unbounded, although f is bounded and holomorphic in an open neighborhood containing the spectrum of A (see [M] and [B-C]).

For polynomially bounded f, one could modify (22.1) by replacing $(w - A)^{-1} x \, dw$ with $(w - A)^{-1} (r - A)^n x \, \frac{dw}{(r-w)^n}$, for n sufficiently large, x in the domain of A^n, r in the resolvent set of A (see [M], [Bo-dL1]). However, this does not include functions of the most interest (particularly in considering the abstract Cauchy problem (0.1) or (22.2) below), such as exponentials and cosines.

When A is unbounded, there are several ways to get bounded operators into the picture and use their functional calculi to define a functional calculus for A. One could apply (22.1) to $(r - A)^{-1}$, for some r in the resolvent set of A; this defines $f(A)$ only for f holomorphic in a neighborhood of infinity (see [Du-S1]), hence is even more restrictive than the requirement of polynomial growth.

One could also attempt to define $f(A)$ by defining a strongly continuous semigroup that it will generate. Unfortunately, even when g maps the spectrum of A into the left half-plane, and $\|(w-A)^{-1}\|$ is $O(\frac{1}{1+|w|})$, $e^{tg}(A)$ may not be bounded. However, it may be shown that $e^{tg}(A)(r-A)^{-1}$ is bounded; more generally, if $\|(w-A)^{-1}\|$ is $O(|w|^{\alpha})$, we will show that $e^{tg}(A)(r-A)^{-m}$, where $m \equiv [\alpha]+2$, is bounded, for all nonnegative t. We will show that this defines a regularized semigroup, so that $g(A)$ may be defined as its generator.

In previous chapters, we have used functional calculus techniques to construct regularized semigroups. In this chapter, we use regularized semigroups to construct a holomorphic functional calculus for operators with polynomially bounded resolvent (see also Chapter XII).

The functional calculus of this chapter may then be used to construct C-regularized semigroups generated by operators with spectrum contained in arbitrary strips and sectors of angle less than $\frac{\pi}{2}$. We may also write down solutions of the abstract Cauchy problem, $u(t,x) = e^{tA}x$, explicitly, using our functional calculus construction of e^{tA}.

By constructing $\cos(itB)$, for an appropriate square root, B, of A, we may similarly write down solutions of the second order abstract Cauchy problem

$$(\frac{d}{dt})^2 u(t) = A\,(u(t))\,(t \in \mathbf{R}),\ u(0) = x,\ u'(0) = y. \qquad (22.2)$$

Throughout this chapter, the sets V and O will be as in Chapter XX.

Definition 22.3. Let $\mathcal{H}_L(V)$ be the set of complex valued functions, f, holomorphic on V such that $\sup\{Re\,(f(z))\,|\,z \in V\} < \infty$. ("L" stands for "left.")

For $\gamma \in \mathbf{R}$, let $\mathcal{H}_\gamma(V)$ be the set of functions, f, holomorphic on V, such that $\sup\{\frac{|f(z)|}{(1+|z|)^\gamma}\,|\,z \in V\} < \infty$. Let $\mathcal{H}_{L,\gamma}(V) \equiv \mathcal{H}_L(V) \cap \mathcal{H}_\gamma(V)$.

Lemma 22.4. *Suppose A is of α-type V. Then there exists O such that $\overline{O} \subseteq V$ and A is of α-type O.*

Theorem 22.5. *Suppose B is of α-type V, $f \in \mathcal{H}_L(V)$ and $m \equiv [\alpha]+2$. Then, for $\lambda \notin V$, $t \geq 0$,*

$$W_f(t) \equiv \frac{1}{2\pi i}\int_{\partial O} e^{tf(w)}(w-B)^{-1}\frac{dw}{(w-\lambda)^m},$$

where O is as in Lemma 22.4, defines a norm continuous $(B-\lambda)^{-m}$-regularized semigroup.

Note that, by the residue theorem, the definition of $W_f(t)$ is independent of O.

134

Definition 22.6. For f, B as in Theorem 22.5, $f(B)$ is defined to be the generator of $\{W_f(t)\}_{t \geq 0}$.

This definition is independent of λ for the following reasons. Temporarily writing $f_\lambda(B)$ for the generator of the $(B - \lambda)^{-m}$-regularized semigroup $\{W_{f,\lambda}(t)\}_{t \geq 0}$ in Theorem 22.5, note that, by Proposition 3.10, for $\lambda_1, \lambda_2 \notin V$, both $f_{\lambda_1}(B)$ and $f_{\lambda_2}(B)$ generate the $(B - \lambda_1)^{-m}(B - \lambda_2)^{-m}$-regularized semigroup

$$\{W_{f,\lambda_1}(t)(B - \lambda_2)^{-m}\}_{t \geq 0} = \{(B - \lambda_1)^{-m} W_{f,\lambda_2}(t)\}_{t \geq 0}$$

(see Lemma 22.35), thus are equal.

For 22.7 through 22.10, let B, α, m, V and O be as in Theorem 22.5.

Proposition 22.7. *If $G \in B(X)$ commutes with all resolvents of B, then $Gf(B) \subseteq f(B)G$, for all $f \in \mathcal{H}_L(V)$.*

Theorem 22.8, Corollary 22.9 and Theorem 22.10 state in what sense the map $f \mapsto f(B)$ is a functional calculus for B. In all these results, there are no growth restrictions on f.

Theorem 22.8. *Suppose B is of α-type V, $\lambda \notin V$, $f, g \in \mathcal{H}_L(V)$ and there exists $\gamma \in \mathbf{R}$ such that $g, fg \in \mathcal{H}_\gamma(V)$. Let $n \equiv \max(0, [\alpha + \gamma] + 2)$. Then $\mathcal{D}(B^n) \subseteq \mathcal{D}(g(B)) \cap \mathcal{D}(f(B)g(B))$ and both $g(B)(B - \lambda)^{-n}$ and $f(B)g(B)(B - \lambda)^{-n}$ are in $B(X)$, with*

$$f(B)g(B)(B - \lambda)^{-k} = \frac{1}{2\pi i} \int_{\partial O} (fg)(w)(w - B)^{-1} \frac{dw}{(w - \lambda)^k},$$

for all $k \geq n$, $k \in \mathbf{N}$.
If $fg \in \mathcal{H}_L(V)$, then $\mathcal{D}(B^n) \subseteq \mathcal{D}((fg)(B))$, and

$$f(B)g(B)(B - \lambda)^{-n} = (fg)(B)(B - \lambda)^{-n}.$$

By choosing $\gamma < -(1 + \alpha)$ and using the fact that the generator of a C-regularized semigroup is automatically closed, we obtain large subsets of domains of our operators, a large class of bounded operators and a type of continuity of the map $f \mapsto f(B)$.

Corollary 22.9. *Suppose $f \in \mathcal{H}_L(V)$, and there exists $\gamma < -(1 + \alpha)$ such that $g \in \mathcal{H}_\gamma(V)$. Then*

(a) $g(B) \in B(X)$, *with*

$$g(B) = \frac{1}{2\pi i} \int_{\partial O} g(w)(w - B)^{-1} \, dw;$$

135

(b) there exists $M_\gamma < \infty$ such that, for all $h \in \mathcal{H}_\gamma(V)$,

$$\|h(B)\| \leq M_\gamma \sup_{w \in V} |\frac{h(z)}{(1+|z|)^\gamma}|; \text{ and}$$

(c) if $fg \in \mathcal{H}_\gamma(V)$, then $Im(g(B)) \subseteq \mathcal{D}(f(B))$, with

$$f(B)(g(B)x) = (fg)(B)x, \forall x \in X.$$

Note that Corollary 22.9(a) implies that Definition 22.6 extends the Riesz-Dunford functional calculus for bounded operators; that is, if B and V are bounded, then Definition 22.6 is equivalent to (22.1).

In the following theorem, there are no growth restrictions on f, fg or $f_1 g$.

Theorem 22.10. Let $f_k(z) \equiv z^k, g_\lambda(z) \equiv \frac{1}{\lambda - z}$.

(a) $f_0(B) = I$.

(b) $g_\lambda(B) = (\lambda - B)^{-1}$, for all $\lambda \notin V$.

(c) If $f, fg \in \mathcal{H}_L(V)$, and there exists $\gamma \in \mathbf{R}$ such that $g \in \mathcal{H}_{L,\gamma}(V)$, then
$$f(B)g(B) \subseteq (fg)(B).$$

(d) If $f_1 g \in \mathcal{H}_L(V), g \in \mathcal{H}_{L,0}(V)$ and $g(B) \in B(X)$, then

$$g(B)B \subseteq Bg(B) \subseteq (f_1 g)(B).$$

(e) If $p \in \mathcal{H}_L(V)$, where p is a polynomial, $p(z) \equiv \sum_{k=0}^n \alpha_k z^k$, then $p(B) = \sum_{k=0}^n \alpha_k B^k$, with $\mathcal{D}(p(B)) = \mathcal{D}(B^k)$.

(f) $f(B) + g(B) \subseteq (f+g)(B)$, for all $f, g \in \mathcal{H}_L(V)$, where $\mathcal{D}(f(B) + g(B)) \equiv \mathcal{D}(f(B)) \cap \mathcal{D}(g(B))$.

(g) $f(B) + \lambda g(B) = (f + \lambda g)(B)$, for all $f \in \mathcal{H}_L(V), g \in \mathcal{H}_{L,0}(V), \lambda \in \mathbf{C}$, when $g(B) \in B(X)$.

In (c), note that, if both $g(B)$ and $f(B)$ are in $B(X)$, then $(fg)(B) \in B(X)$ and $(fg)(B) = f(B)g(B)$, the usual algebra homomorphism property.

Some of the previous results may be summarized by introducing a generalization of an $\mathcal{F}$-functional calculus, what we will call a *C-regularized $\mathcal{F}$-functional calculus*.

136

Definition 22.11. Suppose A is an operator with nonempty C-resolvent (see Definition 3.6), $\mathcal{F}$ is a Banach algebra of complex-valued functions on a subset of the complex plane containing $f_0(z) \equiv 1$ and $g_\lambda(z) \equiv \frac{1}{\lambda - z}$, for some $\lambda \in \rho_C(A)$, and C is a bounded, injective operator that commutes with A. By a *C-regularized $\mathcal{F}$-functional calculus for A* we will mean a continuous linear map, Λ, from $\mathcal{F}$ into $B(X)$ such that

(1) $\Lambda(f_0) = C$;

(2) If $g_\lambda \in \mathcal{F}$, then $\lambda \in \rho_C(A)$ and $\Lambda(g_\lambda) = (\lambda - A)^{-1}C$; and

(3) $\Lambda(fg)C = \Lambda(f)\Lambda(g)$, for all $f, g \in \mathcal{F}$.

Note that, when $C = I$, this is the definition of a $\mathcal{F}$-functional calculus for A. Intuitively, $\Lambda(f) = f(A)C$. This may also be considered a generalization of a C-regularized semigroup, $\{W(t)\}$, generated by A; the operator $W(t)$ may be thought of as $e^{tA}C$.

As with C-regularized semigroups, the idea is that one may have unbounded operators, but the "smoothing" operator, C, provides a uniform control over the unboundedness.

Corollary 22.12. *Suppose B, α, m, V and O are as in Theorem 22.5 and $f(B)$ is as in Definition 22.6. Then, for all $\lambda \notin V$, there exists a $(B - \lambda)^{-m}$-regularized $H^\infty(V)$-functional calculus for B, defined by*

$$\Lambda(f) \equiv f(B)(\lambda - B)^{-m} \equiv \frac{1}{2\pi i} \int_{\partial O} f(w)(w - B)^{-1} \frac{dw}{(w - \lambda)^m}.$$

Iterating the proof of Theorem 22.8 gives us the following.

Corollary 22.13. *Suppose $\{f_i\}_{i=1}^N \subseteq \mathcal{H}_L(V)$, $\prod_{i=1}^j f_i \in \mathcal{H}_\gamma(V)$, for all $j \leq N$, and B, α, λ and n are as in Theorem 22.8. Then $\mathcal{D}(B^n) \subseteq \mathcal{D}\left(\prod_{i=1}^N f_i(B)\right)$ and*

$$\left(\prod_{i=1}^N f_i(B)\right)(B - \lambda)^{-k} = \frac{1}{2\pi i} \int_{\partial O} \left(\prod_{i=1}^N f_i(w)\right)(w - B)^{-1} \frac{dw}{(w - \lambda)^k},$$

for all $k \geq n$, $k \in \mathbf{N}$.

The following theorem shows how we may use our construction to determine spectrum.

Theorem 22.14. *Suppose B is of α-type V, $\mathcal{D}(B)$ is dense, $f \in \mathcal{H}_L(V)$ and there exists $\gamma \in \mathbf{R}$ such that $\left(\frac{1}{f}\right) \in \mathcal{H}_{L,\gamma}(V)$. Then the following are equivalent.*

(a) $\left(\frac{1}{f}\right)(B) \in B(X)$.

(b) $0 \in \rho(f(B))$.

137

Then $(f(B))^{-1} = \left(\frac{1}{f}\right)(B)$.

We then have spectral mapping theorems for functions, f, such that $\frac{1}{f(w)}$ decreases rapidly enough as $|w|$ goes to infinity.

Theorem 22.15 (Spectral mapping theorem). *Suppose B is of α-type V, $\mathcal{D}(B)$ is dense, $f \in \mathcal{H}_L(V)$ and there exists $\gamma < -(1+\alpha)$ such that $\left(\frac{1}{f}\right) \in \mathcal{H}_\gamma(V)$. Then*

$$sp(f(B)) \subseteq f(V).$$

Note that a spectral mapping theorem for arbitrary $f \in H^\infty(V)$ is impossible, since $f(B)$ may be unbounded (see [M] and [B-C]), while if $sp(f(B))$ equalled $f(sp(B))$, $f(B)$ would be bounded since its spectrum is contained in a bounded set.

Thus, at least for $\alpha = -1$, the choice of γ in Theorem 22.15 is best possible. For arbitrary α, the following example shows that, at least for $\alpha \leq 0$, even when $\mathcal{D}(B)$ is dense, spectral mapping theorems may fail if γ is greater than $-(1+\alpha)$; our example is a very natural one, involving *fractional powers* of an operator.

Example 22.16. Suppose $0 \geq \alpha > -1$. This is an example of an operator, B, of α-type $\epsilon + S_{\frac{\pi}{2}}$, such that $sp(B^r) = \mathbf{C}$, whenever $0 < r < (1+\alpha)$; note that $B^r = f_r(B)$, where $\frac{1}{f_r} \in H_{-r}(\epsilon + S_{\frac{\pi}{2}})$, $-r > -(1+\alpha)$, and $f_r(sp(B)) \subseteq S_{\frac{\pi}{2}}$, thus the spectral mapping theorem fails, even for fractional powers.

Define, on $X \times X$,

$$B \equiv \begin{bmatrix} A & A^{2+\alpha} \\ 0 & A \end{bmatrix},$$

$$\mathcal{D}(B) \equiv \{(x_1, x_2) | \, x_1 \in \mathcal{D}(A), A^{-(1+\alpha)}x_1 + x_2 \in \mathcal{D}(A^{2+\alpha})\},$$

where A generates an exponentially decaying strongly continuous holomorphic semigroup on X. Then $(-B)$ is of α-type $(\epsilon + S_{\frac{\pi}{2}})$, for some $\epsilon > 0$, with

$$(\lambda - B)^{-1} = \begin{bmatrix} (\lambda - A)^{-1} & A^{2+\alpha}(\lambda - A)^{-2} \\ 0 & (\lambda - A)^{-1} \end{bmatrix}, \forall \lambda \in \rho(A).$$

Following the construction of Definition 22.6, with $\lambda = 0$, we obtain

$$B^r = \begin{bmatrix} A^r & rA^{(1+\alpha+r)} \\ 0 & A^r \end{bmatrix},$$

$$\mathcal{D}(B^r) = \{(x_1, x_2)| \, x_1 \in \mathcal{D}(A^r), \, A^{-(1+\alpha)}x_1 + rx_2 \in \mathcal{D}(A^{(1+\alpha+r)})\}.$$

For $\lambda \in \rho(A^r)$, on the dense subspace $X \times \mathcal{D}(A^{(1+\alpha-r)})$, one may show that

$$(\lambda - B^r)^{-1} = \begin{bmatrix} (\lambda - A^r)^{-1} & A^{(1+\alpha+r)}(\lambda - A^r)^{-2} \\ 0 & (\lambda - A^r)^{-1} \end{bmatrix},$$

which is an unbounded operator when $0 < r < (1 + \alpha)$. For $\lambda \notin \rho(A^r)$, it is not hard to see that $(\lambda - B^r)$ is not surjective. Thus, $\mathrm{sp}(B^r) = \mathrm{sp}(f_r(B)) = \mathbf{C}$, although $\mathrm{sp}(B) \subseteq \mathrm{sp}(\epsilon + S_{\frac{\pi}{2}})$, and $f_r(\epsilon + S_{\frac{\pi}{2}}) \subseteq S_{\frac{\pi}{2}}$.

Note that, for $r > (1 + \alpha)$, Theorem 22.15 implies that $sp(B^r) \subseteq V^r$, when B is of α-type V and $(-\infty, 0]$ is disjoint from V.

It is natural to ask if Theorem 22.8 could be used to define $f(B)$, that is, define $f(B)$ to be the closure of the operator, defined on $\mathcal{D}(B^n)$, for n sufficiently large, by

$$f(B)x \equiv \frac{1}{2\pi i} \int_{\partial O} f(w)(w - B)^{-1}(B - \lambda)^n x \, \frac{dw}{(w - \lambda)^n}, \qquad (22.17)$$

for $\lambda \notin V, x \in \mathcal{D}(B^n)$. This is the same as asking when $\mathcal{D}(B^n)$ is a core for $f(B)$. The following theorem gives a sufficient condition for this to occur. This includes the hypotheses of Theorem 22.15 (Corollary 22.19). Clearly it would be impossible to use (22.17) when $f \in \mathcal{H}_L(V)$ is not polynomially bounded. Theorem 22.10(a) implies that, when $\mathcal{D}(B)$ is not dense, we also would be unable to use (22.17) to define $f(B)$ equivalently, even when f is polynomially bounded.

Theorem 22.18. *Suppose $\mathcal{D}(B)$ is dense, B is of α-type V, there exists $\gamma \in \mathbf{R}$ such that $f \in \mathcal{H}_{L,\gamma}(V)$ and $\rho(f(B))$ is nonempty. Then $\mathcal{D}(B^n)$ is a core for $f(B)$, for all $n \geq [\gamma + \alpha] + 2$.*

Corollary 22.19. *Suppose $\mathcal{D}(B)$ is dense, B is of α- type V and $\left(\frac{1}{f}\right) \in \mathcal{H}_\gamma(V)$, for some $\gamma < -(1 + \alpha)$. Then $\mathcal{D}(B^k)$ is a core for $f(B)$, for all $k \geq [\gamma] + 1$.*

Remark 22.20. In general, the question of when $\mathcal{D}(B^n)$ is a core for $f(B)$, when f is polynomially bounded, is a special case of the following open question. If $Im(C)$ is dense and A generates a C-regularized semigroup, is $C(\mathcal{D}(A))$ a core for A (see [Dav-P])?

For our examples, let LHP, RHP, S_θ and H_ϵ be as in Chapter XX; let $V_\epsilon \equiv \{z \in \mathbf{C}||Re(z)| < \epsilon\}$.

Example 22.21: FRACTIONAL POWERS, IMAGINARY POWERS AND LOGARITHMS

For arbitrary $\alpha \geq -1$, if B is of α-type S_π, we may define, for $Re(z) \geq 0$, B^z, and define $\log(B)$ as $-ig(B)$, where $g(z) \equiv i\log(z)$, with Definition 22.6.

In particular, for any real t, since $z \mapsto z^{it}$ is bounded on S_π, we may also use Definition 22.6 to define the imaginary powers of B, B^{it}. In fact, it is clear from the definition that $i\log(B)$ is defined as the generator of the B^{-m}-group $\{B^{it}B^{-m}\}_{t\in\mathbf{R}}$ ($m \equiv [\alpha] + 2$).

Traditionally, when B is of (-1)-type S_ω, $\omega < \pi$, the imaginary powers of B are defined by taking the closure of

$$B^{it}x \equiv \int_{\partial S_{\phi,\epsilon}} w^{it}(w - B)^{-1}B^2 x \, \frac{dw}{2\pi i w^2}, \; x \in \mathcal{D}(B^2), \qquad (*)$$

where $S_{\phi,\epsilon} \equiv S_\phi \cap \{z \in \mathbf{C} | |z| > \epsilon\}, \omega < \phi < \pi$ and ϵ is sufficiently small.

By Proposition 3.10, if $\{B^{it}\}_{t\in\mathbf{R}}$ is a strongly continuous group, then its generator is $i\log(B)$, as defined above.

On a Hilbert space, when B is of (-1)-type $S_\omega, \omega < \pi$, $\{B^{it}\}_{t\in\mathbf{R}}$ is a strongly continuous group if and only if there exists an H^∞ functional calculus for B (see [M]).

On a general Banach space, $\{B^{it}\}_{t\in\mathbf{R}}$ being a strongly continuous group has many applications to partial differential equations and operator theory; see [Do-V1, 2], [Pr-Soh], [Y1, 2].

Example 22.22: EXPONENTIALS OF OPERATORS WITH SPECTRUM IN A LEFT HALF-PLANE. Suppose B is of α-type $c + LHP$, for some real c and $m \equiv [\alpha] + 2$. Then by Theorem 22.10(e) (see Definition 22.6), B generates an exponentially bounded $(k - B)^{-m}$-regularized semigroup, for all $k > c$.

Since the map $z \mapsto e^{tz}$ is bounded on $c + LHP$, for all $t \geq 0$, the unbounded operator e^{tB}, yielding the solutions $u(t,x) \equiv e^{tB}x$, of the abstract Cauchy problem (0.1), may be defined directly, by Definition 22.6.

A consequence of Example 22.22 is that an operator with polynomially bounded resolvent outside a vertical strip will generate a C-regularized group. What is more surprising is that the same is true when the spectrum is contained in a *horizontal* strip (see Example 22.24).

Example 22.23: EXPONENTIALS OF OPERATORS WITH REAL SPECTRUM . Suppose the spectrum of B is contained in the real line and B is of α-type H_ϵ, for all $\epsilon > 0$.

For any $s \in \mathbf{R}$, we may define e^{sB}, and hence the solutions of the abstract Cauchy problem (0.1), directly by Definition 22.6, since the map $z \mapsto -e^{sz}$ is in $\mathcal{H}_L(H_{\frac{\pi}{2s}})$.

Note that here we are defining $f(B)$, where f is not polynomially bounded on the spectrum of B.

Example 22.24: EXPONENTIALS AND C-REGULARIZED SEMIGROUPS, SPECTRUM IN A HORIZONTAL STRIP.

More generally, if B is of α-type H_ϵ, for some $\epsilon(0 < \epsilon < 1)$ (this is equivalent to iB generating an exponentially bounded $(i - B)^{-k}$-regularized group, for some $k \in \mathbf{N}$) we may write local solutions of the abstract Cauchy problem down directly, by defining e^{sB}, for $s < \frac{\pi}{2\epsilon}$, with Definition 22.6.

We may also produce global solutions by using Definition 22.6 to define a C-regularized group generated by B. For $t \geq 0$, let $f_t(z) \equiv e^{tz}e^{-z^2}$. Then by Corollary 22.9, $f_t(B) \in B(X)$ and $Bf_t(B) = f_t(B)B = \frac{d}{dt}f_t(B)$, for all $t \geq 0$. This implies that $\{f_t(B)\}_{t \in \mathbf{R}}$ is an $f_0(B)$-regularized group generated by an extension of B (see Theorem 3.7). Since $\rho(B)$ is nonempty, B itself is the generator (see Proposition 3.9).

It may be shown that $f_0(B) = e^{-B^2}$ Again using Corollary 22.9, it may be shown that $Im(e^{-\epsilon B^4}) \subseteq Im(e^{-B^2})$, for all $\epsilon > 0$, which implies that the set of initial data for which the abstract Cauchy problem has a solution is dense, whenever $\mathcal{D}(B)$ is dense.

It is also not hard to show that the C-regularized group generated by B extends to an entire group.

Example 22.25: EXPONENTIALS OF OPERATORS WITH SPECTRUM IN LEFT OR RIGHT SECTORS.

It is well known that B is of α-type $c - S_\theta$ (a left sector), for some $c \in \mathbf{R}, \theta < \frac{\pi}{2}, \alpha \geq -1$, if and only if B generates an exponentially bounded holomorphic k-times integrated semigroup, for some $k \in \mathbf{N}$ (see Chapter XXI). This means that B generates an exponentially bounded holomorphic $(\lambda - B)^{-k}$-regularized semigroup (see Chapter XVIII).

When B is of α-type $c + S_\theta$ (a right sector), for some $c \in \mathbf{R}, \theta < \frac{\pi}{2}$, we may, analogously to Example 22.24, define a C-regularized semigroup, $\{f_t(B)\}_{t \geq 0}$, generated by B, with Definition 22.6, by letting $f_t(z) \equiv e^{tz}e^{-(z-c)^r}$, for some r such that $1 < r < \frac{\pi}{2\theta}$, since f_t will then decay exponentially on $c + S_\theta$.

This may be thought of as a result about *reversibility* of solutions of the abstract Cauchy problem, when sufficiently many of those solutions are analytic in a sector.

Example 22.26: COSINES, SPECTRUM IN VERTICAL STRIP.

If B is of α-type V_ϵ, for some $\epsilon > 0$, we may define $\cos(itB)$ and $\sin(itB)$ directly by Definition 22.6, since the maps $z \mapsto \cos(itz)$ and $z \mapsto \sin(itz)$ are bounded on V_ϵ, for all $t \in \mathbf{R}$. If $m \equiv [\alpha] + 2$, then Theorem 22.8 implies that (22.2), with $A = B^2$, has an exponentially bounded solution, for all initial data in $\mathcal{D}(B^{m+2})$, given by $u(t) = \cos(itB)x + \int_0^t \cos(isB)y \, ds$. In fact, $\{\cos(itB)(\lambda - B)^{-m}\}_{t \in \mathbf{R}}$ will be an exponentially bounded $(\lambda - B)^{-m}$-regularized cosine family generated by B^2, when $|Re(\lambda)| > \epsilon$.

Example 22.27: COSINES, SPECTRUM IN HORIZONTAL STRIP. For operators, B, with spectrum contained in a horizontal strip H_ϵ, as in Example 22.24, we may also define $\cos(itB)$ directly, for sufficiently small $|t|$, and hence treat the second order abstract Cauchy problem (22.2), with $A = B^2$. Or we may define a C-regularized cosine family, $\{g_t(B)\}_{t \in \mathbf{R}}$, where $g_t(z) \equiv \cos(itz)e^{-z^2}$; as in 22.26, this will produce solutions of (22.2), for all initial data in a dense set, when $\mathcal{D}(B)$ is dense.

As in Example 22.23, note that the map $z \mapsto \cos(itz)$ is not polynomially bounded on the spectrum of B.

Example 22.28: COSINES, SPECTRUM IN SECTOR. When B is of α-type $c + S_\theta$, for some real $c, \theta < \frac{\pi}{2}$, as in Example 22.25, we may define a C-regularized cosine family generated by $A = B^2$, $\{h_t(A)\}_{t \in \mathbf{R}}$, where $h_t(z) \equiv \cos(itz)e^{-(z-c)^r}$, where $1 < r < \frac{\pi}{2\theta}$, since h_t will decay exponentially on $c + S_\theta$, giving us solutions of (22.2), as in Example 22.26.

It is clear that we may replace $c + S_\theta$ by a rotation, $\lambda(c + S_\theta)$, for any complex λ, by replacing $e^{-(z-c)^r}$ with $e^{-[\overline{\lambda}(z-c)]^r}$

Again, the map $z \mapsto \cos(itz)$ is not polynomially bounded on the spectrum of B.

Example 22.29: COSINES, SPECTRUM IN SECTOR. When A is of α-type λS_θ, for some complex $\lambda, \theta < \pi$, we may define a C-regularized cosine family generated by A, $\{h_t(A)\}_{t \in \mathbf{R}}$, where $h_t(z) \equiv \cos(it\sqrt{z})e^{-(\sqrt{\lambda z})^r}$, where $\sqrt{z}$ is chosen so as to map S_π into RHP, and $1 < r < \frac{\pi}{2\theta}$, since h_t will decay exponentially on λS_θ, giving us solutions of (22.2), as in Example 22.26.

Example 22.30: THE LAPLACIAN. As a more specific example, we will apply our construction to $B \equiv -\triangle$, on $L^p(\mathbf{R}^n), 1 \le p \le \infty$ (for $\mathcal{D}(B)$ see Example 22.2(2)).

(a) It is known (see [Ri1]) that, for $1 < p < \infty$, $n = 1$, there exists a continuous $H^\infty(S_\theta)$-functional calculus for B, for all $\theta > 0$. Thus $f(B)$ is defined and bounded, for all $f \in H^\infty(S_\theta), \theta > 0$, when $1 < p < \infty, n = 1$. For $1 \le p \le \infty$ and arbitrary n, our construction defines $f(B)$, for $f \in H^\infty(S_\theta - 1)$, however, $f(B)$ may not be bounded. Since B is of (-1)-type $(S_\theta - 1)$, $f(B)(1 - B)^{-1}$ is bounded.

Another way of saying this is that there exists a $(1 - B)^{-1}$-regularized $H^\infty(S_\theta - 1)$-functional calculus for B (see Definition 22.11).

The function $z \mapsto e^{-tz} \in H^\infty(S_\theta)$, for all $\theta \leq \frac{\pi}{2}, t \geq 0$, so this fits into the construction in [Ri1], for $1 < p < \infty, n = 1$. In fact, it is well known that $-B$ generates a bounded holomorphic strongly continuous semigroup, for $1 \leq p < \infty$. For $p = \infty$, $-B$ does not generate a strongly continuous semigroup, but our construction guarantees that it generates a $(1 + B)^{-1}$-regularized semigroup. This enables us to treat the *heat equation*

$$u'(t) = (\Delta u)(t), (t \geq 0)\, u(0) = f,$$

for $f \in L^p(\mathbf{R}^n), 1 \leq p \leq \infty$.

(b) The operator B is not of (-1)-type H_ϵ, for any $\epsilon > 0$. However, there exists n such that B is of n-type H_ϵ, for all $\epsilon > 0$. Thus, for $f \in H^\infty(H_\epsilon)$, we may construct $f(B)$ such that $f(B)(1 + B)^{-(n+2)}$ is bounded.

Another way of saying this is that there exists a $(1+B)^{-(n+2)}$-regularized $H^\infty(H_\epsilon)$-functional calculus for B, for all $\epsilon > 0$ (see Definition 22.11).

This is also saying that $\pm iB$ is as in Example 22.22 so that we may treat the *Schrödinger equation*

$$u'(t) = (i\Delta u)(t)\,(t \in \mathbf{R}), u(0) = f,$$

for $f \in L^p(\mathbf{R}^n), 1 \leq p \leq \infty$.

(c) The operator B also satisfies the hypotheses of Example 22.25. This allows us to treat the *backwards heat equation*,

$$u'(t) + (\Delta u)(t) = 0, u(0) = f,$$

on $L^p(\mathbf{R}^n), 1 \leq p \leq \infty$; we may write down the solution as $u(t) = f_t(B)g$, when $f = e^{-(B-c)^r} g$ (see Example 22.25); this is a dense set, f, of initial data.

(d) The fractional power $B^{\frac{1}{2}}$ may be constructed so as to have the same spectral behaviour as B (this may be seen by using the functional calculus construction for commuting generators of bounded strongly continuous groups—see Chapter XII). Thus, as in Example 22.27 or 22.28, we may treat the *Cauchy problem for the Laplace equation*

$$u''(t) + (\Delta u)(t) = 0\,(t \geq 0), u(0) = f, u'(0) = g,$$

on $L^p(\mathbf{R}^n), 1 \leq p \leq \infty$.

(e) As in (d), since $\Delta = (iB^{\frac{1}{2}})^2$, Example 22.26 may be applied to the *wave equation*

$$u''(t) = (\Delta u)(t)\,(t \geq 0), u(0) = f, u'(0) = g,$$

on $L^p(\mathbf{R}^n), 1 \le p \le \infty$.

Example 22.31. In [Bel, 2], operators, B, of α-type $V_a \equiv \{x + iy \mid x < |y|^a\}$ are considered, where $0 < a < 1$. In effect, what is being constructed there is a C-regularized semigroup generated by B. For $\epsilon > 0$, let $g_{\epsilon,t}(z) \equiv e^{tz}e^{-\epsilon(-z)^b}$, for $a < b < 1$. Then, as in the previous examples in this chapter, $\{g_{\epsilon,t}(B)\}$ is an $e^{-\epsilon(-B)^b}$-regularized semigroup generated by B.

Example 22.32. A MODIFIED HEAT EQUATION. Consider

$$\frac{\partial}{\partial t}u(t,x) = (\frac{\partial}{\partial x})^2 u(t,x) \, (t \ge 0, x \in \mathbf{R}), \, u(t,0) = f(t). \qquad (22.33)$$

Instead of specifying an initial distribution, we are describing the change of temperature at $x = 0$, as time progresses.

A natural choice for X is $C_0[0,\infty)$, the set of all continuous $f : [0,\infty) \to \mathbf{C}$ such that $\lim_{t\to\infty} f(t) = 0$. Let $Af \equiv f'$, on X, with maximal domain. Then (22.33) becomes (22.2) (with x replacing t). Since A generates a bounded strongly continuous semigroup, it has a square root, B, such that B is of (-1)-type $(i - iS_{\frac{\pi}{3}})$. Thus we may use Example 22.28 to deal with (22.33).

We must begin our construction of $f(B)$ with the constant function $f_0(w) \equiv 1$, verifying that $f_0(B)(B-\lambda)^{-m} = (B-\lambda)^{-m}$. This is the same as showing that $W_f(0)$, from Theorem 22.5, equals $(B - \lambda)^{-m}$.

Lemma 22.34. *Suppose B, λ, m, V and O are as in Theorem 22.5. Then, for all $k \ge m$,*

$$(B - \lambda)^{-k} = \frac{1}{2\pi i}\int_{\partial O}(w - B)^{-1}\frac{dw}{(w-\lambda)^k}.$$

Proof: Let D be a disc containing λ, contained in the complement of $\overline{O}$. Let $\Gamma \equiv \partial O \cup \partial D$. Our hypotheses on V imply that there exist complex a, b such that $a + b[0,\infty)$ is contained in the complement of $(V \cup D)$. Let $g(z) \equiv [\overline{b}(z - a)]^{\frac{1}{3}}$, where the fractional power is chosen so that $|\arg(g(w))| < \frac{\pi}{3}$, for all $w \in V$. For $t > 0$, define

$$T(t) \equiv \frac{1}{2\pi i}\int_{\Gamma}e^{-tg(w)}(w - B)^{-1}\,dw.$$

By Cauchy's theorem,

$$2\pi i T(t) = \int_{\partial O}e^{-tg(w)}(w - B)^{-1}\,dw.$$

144

Thus, for x in $\mathcal{D}(B)$,

$$2\pi i T(t)x = \int_{\partial O} e^{-tg(w)}(w - B + B - \lambda)(w - B)^{-1}x \frac{dw}{(w - \lambda)}$$

$$= \left(\int_{\partial O} e^{-tg(w)} \frac{dw}{(w - \lambda)} \right) x$$

$$+ (B - \lambda) \int_{\partial O} e^{-tg(w)}(w - B)^{-1}x \frac{dw}{(w - \lambda)}$$

$$= (B - \lambda) \int_{\partial O} e^{-tg(w)}(w - B)^{-1}x \frac{dw}{(w - \lambda)},$$

by a calculus of residues argument, since B is closed; continuing this argument k times gives

$$2\pi i T(t)(B - \lambda)^{-k} = \int_{\partial O} e^{-tg(w)}(w - B)^{-1} \frac{dw}{(w - \lambda)^k}.$$

By dominated convergence, $\lim_{t \to 0} T(t)(B - \lambda)^{-k}$ exists and equals $\int_{\partial O} (w - B)^{-1} \frac{dw}{(w - \lambda)^k}$. Hence, it is sufficient to show the following:

$$\lim_{t \to 0} T(t)x = x, \forall x \in \mathcal{D}(B^m). \tag{*}$$

145

The calculation follows, for $x \in \mathcal{D}(B^m)$:

$$2\pi i \left(T(t)x - e^{-tg(\lambda)}x \right) = \int_\Gamma \left[e^{-tg(w)}(w-B)^{-1}x - \frac{e^{-tg(w)}}{(w-\lambda)}x \right] dw$$

$$= \int_\Gamma e^{-tg(w)} \left[(w-\lambda+B-B)(w-B)^{-1}x - x \right] \frac{dw}{(w-\lambda)}$$

$$= \int_\Gamma e^{-tg(w)}(w-B)^{-1}(B-\lambda)x \frac{dw}{(w-\lambda)}$$

$$= \int_\Gamma e^{-tg(w)}(w-\lambda+B-B)(w-B)^{-1}(B-\lambda)x \frac{dw}{(w-\lambda)^2}$$

$$= \left[\int_\Gamma e^{-tg(w)} \frac{dw}{(w-\lambda)^2} \right] (B-\lambda)x$$

$$+ \int_\Gamma e^{-tg(w)}(w-B)^{-1}(B-\lambda)^2 x \frac{dw}{(w-\lambda)^2}$$

$$= \dots$$

$$= 2\pi i \sum_{j=1}^{m-1} \frac{1}{j!} \left(\frac{d}{dz} \right)^j \left(e^{-tg(z)} \right) \big|_{z=\lambda} (B-\lambda)^j x$$

$$+ \int_\Gamma e^{-tg(w)}(w-B)^{-1}(B-\lambda)^m x \frac{dw}{(w-\lambda)^m}.$$

Each term in the sum is divisible by t, hence goes to zero, as t does. The integral converges to zero, as t does, by dominated convergence and the residue theorem. This establishes $(*)$ and the Lemma. ∎

Lemma 22.35. *Suppose* B, λ, m, V *and* O *are as in Theorem 22.5, and* h *and* g *are bounded and holomorphic on* V.

(a)

$$\left[\int_{\partial O} h(w)(w-B)^{-1} \frac{dw}{(w-\lambda)^m} \right] \left[\int_{\partial O} g(z)(z-B)^{-1} \frac{dz}{(z-\lambda)^m} \right]$$

$$= (2\pi i) \int_{\partial O} (hg)(z)(z-B)^{-1} \frac{dz}{(z-\lambda)^{2m}}.$$

(b) For all $x \in X$, $\int_{\partial O} h(w)(w-B)^{-1}x \frac{dw}{(w-\lambda)^{m+1}} \in \mathcal{D}(B)$, with

$$(B-\lambda) \int_{\partial O} h(w)(w-B)^{-1}x \frac{dw}{(w-\lambda)^{m+1}} = \int_{\partial O} h(w)(w-B)^{-1}x \frac{dw}{(w-\lambda)^m}.$$

146

Proof: (a) Let $\Gamma_1 \equiv \partial O$. There exists another smooth curve, Γ_2, contained in $(V - \overline{O})$. We calculate as follows, using the residue theorem.

$$\left[\int_{\partial O} h(w)(w-B)^{-1} \frac{dw}{(w-\lambda)^m} \right] \left[\int_{\partial O} g(z)(z-B)^{-1} \frac{dz}{(z-\lambda)^m} \right]$$

$$= \int_{\Gamma_2} \int_{\Gamma_1} h(w)g(z)\frac{1}{(w-z)} \left((z-B)^{-1} - (w-B)^{-1} \right) \frac{dz}{(z-\lambda)^m} \frac{dw}{(w-\lambda)^m}$$

$$= \int_{\Gamma_1} \left(\int_{\Gamma_2} \frac{h(w)}{(w-\lambda)^m} \frac{dw}{(w-z)} \right) g(z)(z-B)^{-1} \frac{dz}{(z-\lambda)^m}$$

$$+ \int_{\Gamma_2} \left(\int_{\Gamma_1} \frac{g(z)}{(z-\lambda)^m} \frac{dz}{(z-w)} \right) h(w)(w-B)^{-1} \frac{dw}{(w-\lambda)^m}$$

$$= \int_{\Gamma_1} \frac{(2\pi i)h(z)}{(z-\lambda)^m} g(z)(z-B)^{-1} \frac{dz}{(z-\lambda)^m}.$$

(b) For $w \in \partial V$,

$$(B-\lambda)(w-B)^{-1}x = (B-w+w-\lambda)(w-B)^{-1}x = (w-\lambda)(w-B)^{-1}x - x,$$

so that $\|\frac{1}{(w-\lambda)^{m+1}}(B - \lambda)(w - B)^{-1}x\|$ is $O\left((1+|w|)^{-2}\right)$. Since B is closed, this implies that $\int_{\partial O} h(w)(w-B)^{-1}x \frac{dw}{(w-\lambda)^{m+1}} \in \mathcal{D}(B)$, with

$$(B-\lambda)\int_{\partial O} h(w)(w-B)^{-1}x \frac{dw}{(w-\lambda)^{m+1}}$$

$$= \int_{\partial O} h(w) \left[(w-\lambda)(w-B)^{-1}x - x \right] \frac{dw}{(w-\lambda)^{m+1}}$$

$$= \int_{\partial O} h(w)(w-B)^{-1}x \frac{dw}{(w-\lambda)^m},$$

by the residue theorem. ∎

Proof of Lemma 22.4: There exists $M < \infty$ such that

$$\|(w-A)^{-1}\| \le M(1+|w|)^\alpha, \ \forall w \notin V.$$

Suppose $w \notin V$ and $|z - w| \le \frac{1}{2M(1+|w|)^\alpha}$. Then $(z-A)^{-1}$ exists and is given by $\sum_{k=0}^{\infty}(w-z)^k(w-A)^{-(k+1)}$, so that

$$\|(z-A)^{-1}\| \le \sum_{k=0}^{\infty} \left[\frac{1}{2M(1+|w|)^\alpha} \right]^k (M(1+|w|)^\alpha)^{(k+1)} = 2M(1+|w|)^\alpha.$$

$$(*)$$

Let $K \equiv \bigcup_{w \notin V} \{z \mid |z - w| \leq \min\left(\frac{1}{2M(1+|w|)^\alpha}, \frac{|w|}{2}\right)\}$. Choose O as in the last chapter, such that $\overline{C - K} \subseteq O$ and $\overline{O} \subseteq V$.

By $(*)$, to show that A is of α-type O, it is sufficient to show that $|[\frac{1+|w|}{1+|z|}]^\alpha]|$ is bounded as $|z|$ goes to ∞. This is clear from the fact that

$$\left|\frac{1+|z|}{1+|w|} - 1\right| = \left|\frac{|z|-|w|}{1+|w|}\right| \leq \frac{|z-w|}{1+|w|} < \frac{1}{2M(1+|w|)^{\alpha+1}},$$

and $\alpha + 1 \geq 0$, by choosing $M > 1$, if necessary. ∎

Proof of Theorem 22.5: By Lemma 22.35(a),

$$W_f(t)W_f(s) = W_f(0)W_f(t+s) \quad (s, t \geq 0).$$

By Lemma 22.34, $W_f(0) = (B - \lambda)^{-m}$. The desired continuity of $t \to W_f(t)$ follows from dominated convergence. ∎

Proof of Proposition 22.7: Suppose $x \in \mathcal{D}(f(B))$ and $t > 0$. Then

$$\frac{1}{t}\left(W_f(t)Gx - (B-\lambda)^{-m}Gx\right) = G\left(\frac{1}{t}\left(W_f(t)x - (B-\lambda)^{-m}x\right)\right);$$

thus, since G is bounded, $\lim_{t\to 0}$ exists, and equals

$$G(B-\lambda)^{-m}f(B)x = (B-\lambda)^{-m}Gf(B)x,$$

so that, by the definition of the generator, $Gx \in \mathcal{D}(f(B))$, with $f(B)Gx = Gf(B)x$. ∎

Proof of Theorem 22.8: (a) First, we will show that $\mathcal{D}(B^n) \subseteq \mathcal{D}((g(B)))$, with

$$g(B)(B-\lambda)^{-k} = \frac{1}{2\pi i}\int_{\partial O} g(w)(w-B)^{-1}\frac{dw}{(w-\lambda)^k}, \qquad (*)$$

for all $k \geq n$.

As in the proof of Theorem 22.5, it may be shown that the $(B-\lambda)^{-(k+m)}$-regularized semigroup $\{W_g(t)(B-\lambda)^{-k}\}_{t\geq 0}$ (generated by $g(B)$— see Proposition 3.10) is given by

$$2\pi i W_g(t)(B-\lambda)^{-k} = \int_{\partial O} e^{tg(w)}(w-B)^{-1}\frac{dw}{(w-\lambda)^{k+m}}.$$

Since $\|g(w)(w-B)^{-1}\|$ is $O((1+|w|)^{\alpha+\gamma})$, we may differentiate as follows, for $x \in X$:

$$2\pi i\frac{d}{dt}W_g(t)(B-\lambda)^{-k}x|_{t=0} = \int_{\partial O} \frac{d}{dt}\left(e^{tg(w)}|_{t=0}\right)(w-B)^{-1}x\frac{dw}{(w-\lambda)^{k+m}}$$

$$= \int_{\partial O}\left[\frac{g(w)}{(w-\lambda)^k}\right](w-B)^{-1}x\frac{dw}{(w-\lambda)^m};$$

by Lemma 22.35(b), this equals

$$(B - \lambda)^{-m} \int_{\partial O} g(w)(w - B)^{-1} x \, \frac{dw}{(w - \lambda)^k},$$

which establishes $(*)$, by the definition of $g(B)$ as the generator of the $(B - \lambda)^{-m}$-regularized semigroup $\{W_g(t)\}_{t \geq 0}$.

For $t \geq 0$, Lemma 22.35(a) and $(*)$ now imply that

$$2\pi i W_f(t) g(B)(B - \lambda)^{-k} = \int_{\partial O} e^{t f(w)} g(w)(w - B)^{-1} \, \frac{dw}{(w - \lambda)^{k+m}},$$

so that, as with the proof of $(*)$, for $x \in X$, we may differentiate:

$$2\pi i \frac{d}{dt} W_f(t) g(B)(B - \lambda)^{-k} x|_{t=0}$$
$$= \int_{\partial O} f(w) g(w)(w - B)^{-1} x \, \frac{dw}{(w - \lambda)^{k+m}}$$
$$= (B - \lambda)^{-m} \int_{\partial O} f(w) g(w)(w - B)^{-1} \, \frac{dw}{(w - \lambda)^k},$$

by Lemma 22.35(b).

This implies that, for all $x \in X$, $g(B)(B - \lambda)^{-k} x \in \mathcal{D}\left(f(B)\right)$, so that $(B - \lambda)^{-k} x \in \mathcal{D}\left(f(B) g(B)\right)$, with

$$f(B) g(B)(B - \lambda)^{-k} x = \frac{1}{2\pi i} \int_{\partial O} (fg)(w)(w - B)^{-1} x \, \frac{dw}{(w - \lambda)^k}.$$

By $(*)$, this equals $(fg)(B)(B - \lambda)^{-k} x$, when $(fg) \in \mathcal{H}_L(V)$. ∎

Proof of Theorem 22.10: (a) $W_{f_0}(t) = \frac{1}{2\pi i} \int_{\partial O} e^t (w - B)^{-1} \frac{dw}{(w-\lambda)^m}$ is clearly differentiable and equal to its own derivative, thus, by Lemma 22.34, $\mathcal{D}(f_0(B)) = X$, with $f_0(B) x \equiv (B - \lambda)^m \frac{d}{dt} W_{f_0}(t) x|_{t=0} = x$, for all $x \in X$.

(b) $W_{g_\lambda}(t) = \frac{1}{2\pi i} \int_{\partial O} e^{t(\lambda - w)^{-1}} (w - B)^{-1} \frac{dw}{(w - \lambda)^m}$, thus we may differentiate as follows:

$$\frac{d}{dt} W_{g_\lambda}(t) = -\frac{1}{2\pi i} \int_{\partial O} e^{t(\lambda - w)^{-1}} (w - B)^{-1} \, \frac{dw}{(w - \lambda)^{m+1}} = (\lambda - B)^{-1} W_{g_\lambda}(t),$$

by Lemma 22.35(b). This implies that $\mathcal{D}(g_\lambda(B)) = X$, with $g_\lambda(B) x \equiv (B - \lambda)^m \frac{d}{dt} W_{g_\lambda}(t) x|_{t=0} = (\lambda - B)^{-1} x$.

149

(c) Let $n \equiv [\alpha + \gamma] + 2$ and fix $\lambda \notin V$. By Theorem 22.8, $g(B)(\lambda - B)^{-n} \in B(X)$.

By the definition of $(fg)(B)$, to show that $f(B)g(B) \subseteq (fg)(B)$, it is sufficient to show that

$$\int_0^t W_{fg}(s)\,(f(B)g(B)x)\,ds = W_{fg}(t)x - W_{fg}(0)x,$$

for all $x \in \mathcal{D}(f(B)g(B)), t \geq 0$.

Since B is closed, this is equivalent to

$$\int_0^t W_{fg}(s)\,(f(B)g(B)(\lambda - B)^{-n}x)\,ds \tag{$*$}$$
$$= W_{fg}(t)(\lambda - B)^{-n}x - W_{fg}(0)(\lambda - B)^{-n}x,$$

for all $x \in \mathcal{D}(f(B)g(B)), t \geq 0$.

Fix $x \in \mathcal{D}(f(B)g(B)), t \geq 0$. By Proposition 22.7, $W_{fg}(s)x$ is in the domain of $(f(B)g(B))$, and $f(B)g(B)W_{fg}(s)x = W_{fg}(s)f(B)g(B)x$, for all $s \geq 0$. Since $g(B)(\lambda - B)^{-n}$ is bounded and $f(B)$ is closed, $f(B)g(B)(\lambda - B)^{-n}$ is closed. Thus,

$$y \equiv \int_0^t W_{fg}(s)x\,ds \in \mathcal{D}\left(f(B)g(B)(\lambda - B)^{-n}\right),$$

with

$$f(B)g(B)(\lambda - B)^{-n}y = \int_0^t W_{fg}(s)\,(f(B)g(B)(\lambda - B)^{-n}x)\,ds.$$

Since $y = \int_{\partial O} \left[\int_0^t e^{s(fg)(w)}\,ds\right](w - B)^{-1}x\,\frac{dw}{(w-\lambda)^m}$, Corollary 22.9 and Lemma 22.35 imply that, for all $r \geq 0$,

$$2\pi i W_f(r)g(B)(\lambda - B)^{-n}y$$
$$= \int_{\partial O} e^{rf(w)}g(w)\left[\int_0^t e^{s(fg)(w)}\,ds\right](w - B)^{-1}x\,\frac{dw}{(w - \lambda)^{2m+n}}.$$

Since the r-derivative of the integrand is an L^1 function of w, uniformly in r, we may differentiate under the integral sign:

$$2\pi i \frac{d}{dr}W_f(r)g(B)(\lambda - B)^{-n}y$$
$$= \int_{\partial O} e^{rf(w)}\left(e^{t(fg)(w)} - 1\right)(w - B)^{-1}x\,\frac{dw}{(w - \lambda)^{2m+n}}.$$

150

By Lemma 22.35(b), this implies that

$$f(B)g(B)(\lambda - B)^{-n}y$$

$$= \frac{1}{2\pi i}(B - \lambda)^m \int_{\partial O} \left(e^{t(fg)(w)} - 1\right)(w - B)^{-1}x \, \frac{dw}{(w - \lambda)^{2m+n}}$$

$$= \left(W_{fg}(t)(\lambda - B)^{-n}x - W_{fg}(0)(\lambda - B)^{-n}x\right),$$

proving $(*)$, as desired.

(d) It is clear from the definition of $g(B)$ that it commutes with all resolvents of B; this implies that $g(B)B \subseteq Bg(B)$. As in (c), to show that $Bg(B) \subseteq (f_1g)(B)$, it is sufficient to show that

$$\int_0^t W_{f_1g}(s)(Bg(B)x) \, ds = W_{f_1g}(t)x - W_{f_1g}(0)x,$$

for all $x \in \mathcal{D}(Bg(B)), t \geq 0$. Since B is closed, this is equivalent to showing that

$$\int_0^t W_{f_1g}(s)\left(B(\lambda - B)^{-1}g(B)x\right) \, ds = (\lambda - B)^{-1}\left(W_{f_1g}(t)x - W_{f_1g}(0)x\right),$$

for $\lambda \notin V$. This follows by a simple integration, after applying Lemma 22.35(b), Theorem 22.10(b) and Corollary 22.9(a) and writing $B(\lambda - B)^{-1}$ as $\lambda(\lambda - B)^{-1} - I$.

(e) Let us write $p(B)_1$ for the operator defined by

$$p(B)_1 x \equiv \sum_{k=0}^n \alpha_k B^k x, \quad \mathcal{D}(p(B)_1) \equiv \mathcal{D}(B^k).$$

We need to show that $p(B)_1 = p(B)$. Note that, since $p(V)$ is contained in a left half-plane, there exists complex λ such that $p(\lambda) \notin p(V)$. This implies that there exist not necessarily distinct λ_i, outside V, such that

$$p(\lambda) - p(B)_1 = \prod_{i=1}^n (\lambda_i - B).$$

Suppose $x \in \mathcal{D}(p(B)_1)$. Then, by Lemma 22.35(b),

$$2\pi i W_p(t)x = \int_{\partial O} e^{tp(w)}(w - B)^{-1}(p(\lambda) - p(B)_1)x \, \frac{dw}{[\prod_{i=1}^n (\lambda_i - w)](\lambda - w)^m},$$

151

so that we may differentiate

$$\frac{d}{dt}W_p(t)x|_{t=0}$$

$$= \int_{\partial O} \frac{p(w)}{\prod_{i=1}^n(\lambda_i - w)}(w - B)^{-1}(p(\lambda) - p(B)_1)x\frac{dw}{2\pi i(\lambda - w)^m}$$

$$= (\lambda - B)^{-m}p(B)_1 x,$$

by Lemmas 22.35(b) and 22.34. This implies that $x \in \mathcal{D}(p(B))$, with $p(B)x = p(B)_1 x$, that is, $p(B)_1 \subseteq p(B)$. Since $\rho(p(B)_1)$ is nonempty, this implies that $p(B)_1 = p(B)$ (see Proposition 3.9).

(f) By Lemma 22.35(a), $\{W_f(t)W_g(t)\}_{t\geq 0} = \{W_{f+g}(t)(B-\lambda)^{-m}\}_{t\geq 0}$, the $(B - \lambda)^{-2m}$-regularized semigroup generated by $(f + g)(B)$. In general, if $\{W_i(t)\}_{t\geq 0}$ is a C_i-regularized semigroup generated by G_i, for $i = 1, 2$, and $W_1(t)W_2(s) = W_2(s)W_1(t)$, for all $s, t \geq 0$, then $\{W_1(t)W_2(t)\}_{t\geq 0}$ is a C_1C_2-regularized semigroup generated by an extension of $(G_1 + G_2)$.

(g) By (f), it is sufficient to show that $\lambda g(B) = (\lambda g)(B)$. This is clear from Definition 22.6. ∎

Proof of Theorem 22.14: (a) $\rightarrow$ (b). By Theorem 22.10(a) and (c), $G \equiv f(B)\left(\frac{1}{f}\right)(B) \subseteq I$. Since $\left(\frac{1}{f}\right)(B) \in B(X)$, G is closed. By Theorem 22.8, since $\frac{1}{f} \in \mathcal{H}_\gamma(V)$, G is densely defined; thus, taking closures on both sides of the inclusion implies that $f(B)\left(\frac{1}{f}\right)(B) \equiv G = I$. Proposition 22.7 implies that $\left(\frac{1}{f}\right)(B)f(B) \subseteq I$.

(b) $\rightarrow$ (a). By Theorem 22.8, there exists $n \in \mathbf{N}$ such that $f(B)\left(\frac{1}{f}\right)(B)(\lambda - B)^{-n} = (\lambda - B)^{-n}(\lambda \notin V)$, so that $\left(\frac{1}{f}\right)(B)(\lambda - B)^{-n} = (f(B))^{-1}(\lambda - B)^{-n}$. Since $\left(\frac{1}{f}\right)(B)$ is closed and $\mathcal{D}(B^n)$ is dense, we may take closures of both sides to conclude that $\left(\frac{1}{f}\right)(B) = (f(B))^{-1} \in B(X)$. ∎

Proof of Theorem 22.15: This follows from Theorem 22.14 and Corollary 22.9(a), since, for $\lambda \notin V$, $\frac{1}{f-\lambda} = \left(\frac{f}{f-\lambda}\right)\left(\frac{1}{f}\right)$ and the first term is bounded on O, for O as in Lemma 22.4. ∎

For Theorem 22.18, we will need the following ([Ka, Problem III-5.19]).

Lemma 22.36. *Suppose* $\mathcal{D} \subseteq \mathcal{D}(A), r \in \rho(A),$ *and* $(r - A)(\mathcal{D})$ *is dense. Then* $\mathcal{D}$ *is a core for* A.

Proof of Theorem 22.18: It is sufficient to assume $0 \in \rho(f(B))$. To see that $(f(B))(\mathcal{D}(B^k))$ is dense, note that, by Theorem 22.8, there exists

152

$m \in \mathbf{N}$ such that

$$(f(B))(B - \lambda)^{-k} \left(\frac{1}{f}\right)(B)(B - \lambda)^{-m} = (B - \lambda)^{-(m+k)} \quad (\lambda \notin V).$$

This implies that $(f(B))\left(\mathcal{D}(B^k)\right)$ contains $\mathcal{D}(B^{m+k})$, which is dense, since $\mathcal{D}(B)$ is dense. By Lemma 22.36, $\mathcal{D}(B^k)$ is a core for $f(B)$. ∎

153

XXIII. SPECTRAL CONDITIONS GUARANTEEING SOLUTIONS OF THE ABSTRACT CAUCHY PROBLEM

It is natural and very common to ask what conditions on the resolvent of A, $(w - A)^{-1}$, will guarantee that the abstract Cauchy problems (0.1) and (22.2) have solutions, for all initial data in dense sets. In this chapter, we characterize subsets, V, of the complex plane, with the property that, whenever A is of α-type V (Definition 20.1) and densely defined, then the abstract Cauchy problem has a solution, for all initial data, x, in a dense set (Theorem 23.6). We give a similar result for (22.2) (Theorem 23.11).

These theorems reduce the operator theoretic question, of when solutions of (0.1) and (22.2) exist, to the complex analytic question of when there exist holomorphic g, on V, such that $z \mapsto e^{tz}g(z)$ is bounded, for all $t \geq 0$ (for (0.1)), or when there exist holomorphic g such that $z \mapsto \cos(itz)g(z)$ is bounded, for all real t (for (22.2)).

Definition 23.1. If V is as in Chapter XX, we will say that V is an *α-spectral dense solution set* if, whenever A is of α-type V and densely defined, then the abstract Cauchy problem (0.1) has a solution, for all initial data in a dense set.

The simplest sufficient condition for being an α-spectral solution set is independent of α. We will see that, in general, this property is not independent of α (see Theorem 23.6, Example 23.7).

Definition 23.2. We will denote by E_V the set of holomorphic functions g on V that decay more rapidly than any exponential, $\{g|\, z \mapsto e^{tz}g(z) \in H^\infty(V)$, for all $t \geq 0\}$.

Theorem 23.3. *Suppose the constant function $f_0(z) \equiv 1$ is the pointwise limit of a bounded sequence from E_V. Then V is an α-spectral dense solution set.*

The sufficient condition of Theorem 23.3 is not necessary, because when an operator is of α-type V, the V is not minimal, that is, we can always find a smaller set, O, such that A is of α-type O (Lemma 23.12). It is necessary and sufficient that there exist sufficiently many $g \in E_O$, whenever O is a subset of V that is asymptotically close enough (Theorem 23.6). The closeness depends on α (see Examples 23.5 and 23.7).

The following definition measures how asymptotically close to V this subset, O, must be.

Definition 23.4. The set of complex numbers O is an *α-interior subset of V* if O is as in Chapter XX, $\overline{O} \subseteq V$ and $d(O, \{z \notin V |\, |z| \leq R\})$ is $O((1 + R)^{-\alpha})$.

154

Example 23.5. Let $V \equiv RHP \equiv \{z \in \mathbf{C} | Re(z) > 0\}$. Then, for all $\theta < \frac{\pi}{2}$, $S_\theta \equiv \{re^{i\phi} | r > 0, |\phi| < \theta\}$ is a (-1)-interior subset of V, but not a 0-interior subset (or an α-interior subset, for any $\alpha > -1$). For any $\epsilon > 0$, $\epsilon + RHP$ is a 0-interior subset of V, but not an α-interior subset, for any $\alpha > 0$. An example of a 1-interior subset of V would be $\{x + iy | x > (1 + |y|)^{-1}\}$.

More generally, for any $\alpha \geq -1$, $O_{\alpha,k}$, from Example 20.2(3), is an α-interior subset of V, for all $k > 0$, and is not a β-interior subset of V, for any $\beta > \alpha$.

As suggested by this example, the following relationship between Definition 23.4 and 20.1 is not hard to see. O is an α-interior subset of V if and only if the operator $Af(z) \equiv zf(z)$, on $L^p(\overline{O}), 1 \leq p \leq \infty$, is an operator of α-type V.

Theorem 23.6. *The following are equivalent.*

(a) *The set V is an α-spectral dense solution set.*

(b) *For all α-interior subsets, O, of V, the constant function $f_0(z) \equiv 1$ is the pointwise limit of a bounded sequence from E_O.*

(c) *For all α-interior subsets, O, of V, E_O is uniformly dense in $H^\infty(O)$.*

Example 23.7. Let $V \equiv RHP$, as in Example 23.5. Then V is a (-1)-spectral dense solution set, but is not a 0-spectral dense solution set.

To see this, note first that any (-1)-interior subset, O, of V is contained in S_θ, for some $\theta < \frac{\pi}{2}$, thus $g_\epsilon(z) \equiv e^{-\epsilon z^r}$, where r is chosen so that $1 < r < \frac{\pi}{2\theta}$, and $\epsilon > 0$ is arbitrary, satisfies Theorem 23.6(b). Thus V is a (-1)-spectral dense solution set, by Theorem 23.6. To see that V is not a 0-spectral dense solution set, consider $A \equiv 1 + \frac{d}{dx}$, on $X \equiv \{f \in C[0,1] | f(0) = 0\}$, $\mathcal{D}(A) \equiv \{f \in X | f' \in X\}$. Then it is not hard to show that A is of 0-type V, since $1 - A$ generates a bounded strongly continuous semigroup, but the abstract Cauchy problem (0.1) has no nontrivial solutions.

Essentially the same proofs give the same results for second order problems (22.2), with e^{tz} replaced by $\cos(itz)$.

Definition 23.8. If V is as in Chapter XXII, we will say that V is a *second order α-spectral dense solution set* if, whenever B is of α-type V and densely defined, then (22.2), with $A = B^2$, has a solution, for all initial data in a dense set.

Definition 23.9. We will denote by C_V the set of all holomorphic functions g on V such that the map $z \mapsto \cos(itz)g(z) \in H^\infty(V)$, for all $t \geq 0$.

Theorem 23.10. *Suppose the constant function $f_0(z) \equiv 1$ is the pointwise limit of a bounded sequence from C_V. Then V is a second order α-spectral dense solution set.*

Theorem 23.11. *The following are equivalent.*

(a) *The set V is a second order α-spectral dense solution set.*

(b) *For all α-interior subsets, O, of V, the constant function $f_0(z) \equiv 1$ is the pointwise limit of a bounded sequence from C_O.*

(c) *For all α-interior subsets, O, of V, C_O is uniformly dense in $H^\infty(O)$.*

The proof of Lemma 22.4 proves the following, which is well known for $\alpha = -1, V = S_\theta$ (see [Balak]).

Lemma 23.12. *Suppose A is of α-type V. Then there exists an α-interior subset, O, of V, such that A is of α-type O.*

Proof of Theorem 23.3: Suppose A is of α-type V. Let $m \equiv [\alpha] + 2$. Let $\{g_k\}_{k=1}^\infty$ be a bounded sequence from E_V that converges to f_0 pointwise. Fix $\lambda \notin V$. For $t \geq 0, x \in X, k \in \mathbf{N}$, let $h_{t,k}(z) \equiv (\lambda - z)^{-(m+1)} e^{tz} g_k(z)$ and let $u_{k,x}(t) \equiv h_{t,k}(A)x$. By Corollary 22.9 and Theorem 22.10(d), $u_{k,x}(t)$ is a continuously differentiable function of t and $\frac{d}{dt} u_{k,x}(t) = A(u_{k,x}(t))$, for all $t \geq 0$. Since $\mathcal{D}(A^{m+1})$ is dense, all that remains is to show that $(\lambda - A)^{-(m+1)} x = \lim_{k \to \infty} u_{k,x}(0)$. This follows by dominated convergence, Corollary 22.9(a) and Theorem 22.10(b). ∎

Proof of Theorem 23.6: (b) → (a). Suppose A is of α-type V and densely defined. By Lemma 23.12, there exists an α-interior subset, O, of V, such that A is of α-type O. Assertion (a) now follows from Theorem 23.3.

(a) → (c). Let O be an α-interior subset of V. Let $(Af)(z) \equiv zf(z)$, on $H^\infty(O)$, with maximal domain. Then A is of α-type V, thus (0.1) has a solution, $t \mapsto u(t,g)$, for all g in a dense set, $\mathcal{D}$. It is clear that u has the form $(u(t,g))(z) = e^{tz} g(z)$, for $t \geq 0, z \in O$. Thus $g \in E_O$, for all $g \in \mathcal{D}$, as desired.

(c) → (b) is obvious. ∎

Example 23.13. For any real $c, \alpha \geq -1$, $V \equiv (c + LHP)$ is an α-spectral dense solution set, since $f_0(z) \equiv 1 \in E_V$. In fact, it is known that an operator of α-type V generates an $[\alpha] + 2$-times integrated semigroup and a $(\lambda - A)^{-([\alpha]+2)}$-regularized semigroup (see Chapter XXI).

Example 23.14. For any $\alpha \geq -1, \epsilon > 0$, the vertical strip $V_\epsilon \equiv \{z \in \mathbf{C} \| |Re(z)| < \epsilon\}$ is a second order α-spectral dense solution set, since $f_0 \in C_{V_\epsilon}$ (see [A-Ke]).

Example 23.15. For any real c, $\alpha \geq -1$, $0 < \theta < \frac{\pi}{2}$, $V \equiv c + S_\theta$, where $S_\theta \equiv \{re^{i\phi}||\phi| < \theta\}$, is both an α-spectral dense solution set and a second order α-spectral dense solution set.

To see this, choose r such that $1 < r < \frac{\pi}{2\theta}$. For any $\epsilon > 0$, let $g_\epsilon(z) \equiv e^{-\epsilon z^r}$. Then it may be shown that, for all $\epsilon > 0$, g_ϵ is in both E_V and C_V. Since $\lim_{\epsilon \to 0} g_\epsilon(z) = 1$, for all $z \in V$, we may apply Theorems 23.6 and 23.11.

As mentioned in Example 22.25, this may be considered a result about reversibility of solutions.

We discussed $S_{\frac{\pi}{2}} = RHP$ in Example 23.7.

Example 23.16. For any $c > 0$, $\alpha \geq -1$, the horizontal strip $H_c \equiv \{z \in \mathbf{C} \,||Im(z)| < c\}$ is both an α-spectral dense solution set and a second order α-spectral dense solution set.

As in the previous example, this may be seen by showing that $g_\epsilon(z) \equiv e^{-\epsilon z^2}$ is in both E_{H_c} and C_{H_c}, for all $\epsilon > 0$.

Example 23.17. For $\theta < \frac{\pi}{4}$, $\alpha \geq -1$, $c \in \mathbf{R}$, the same choice of g_ϵ as in the previous example shows that the double sector $(c + S_\theta) \cup (c - S_\theta)$ is both an α-spectral dense solution set and a second order α-spectral dense solution set.

Example 23.18. In [Bel, 2], operators of α-type $V_a \equiv \{x + iy| x < |y|^a\}$ are considered. In the language of this chapter, it is shown there that V_a is an α spectral dense solution set, whenever $0 < a < 1$. This follows from Theorem 23.6, by considering $g_\epsilon(z) \equiv e^{-\epsilon(-z)^b}$, for $a < b < 1$, and showing that this is in E_{V_a}, for all $\epsilon > 0$; in fact, this is the key to the construction in [Bel].

In [Bel] it is also shown that $V \equiv \{x + iy\,|\,x < |y|[\log(1 + |y|)]^{-1}\}$ is not a (-1)-spectral dense solution set. This also follows from Theorem 23.6, after showing that there exists no g in E_O.

XXIV. POLYNOMIALS OF GENERATORS

In this chapter, we prove a type mapping theorem (see Chapter XX) for polynomials; that is, we show that, if A is of n-type Ω, then $p(A)$ is of n-type $p(\Omega)$, for any polynomial p. When p maps Ω into an appropriate sector, this allows us to use the results of Chapter XXI to conclude that $p(A)$ generates an exponentially bounded holomorphic k-times integrated semigroup.

Our definition of type Ω, in Chapter XX, involves only growth conditions on the resolvent at infinity. To obtain a bounded strongly continuous holomorphic semigroup, we also need growth conditions at zero (see Lemma 24.12).

If p is a polynomial, $p(x) \equiv \sum_{k=0}^{N} c_k x^k$; then we define

$$p(A) \equiv \sum_{k=0}^{N} c_k A^k, \ \mathcal{D}(p(A)) \equiv \mathcal{D}(A^N).$$

Theorem 24.1. *Suppose A is of n-type Ω and p is a polynomial. Then $p(A)$ is of n-type $p(\Omega)$.*

Theorem 24.2. *Suppose A is of $(k-1)$-type Ω, $\frac{\pi}{2} > \theta > 0$, and q is a polynomial such that $q(\Omega) \subseteq S_\theta$. Then*

(a) *the operator $-q(A)$ generates an exponentially bounded holomorphic $(k+1)$-times integrated semigroup of angle $(\frac{\pi}{2} - \theta)$;*

(b) *if $\mathcal{D}(q(A))$ is dense, then $-q(A)$ generates an exponentially bounded holomorphic k-times integrated semigroup of angle $(\frac{\pi}{2} - \theta)$; and*

(c) *if $k = 0$, $\mathcal{D}(q(A))$ is dense, and $q(0) = 0$, then $-q(A)$ generates a bounded holomorphic strongly continuous semigroup of angle $(\frac{\pi}{2} - \theta)$.*

Corollary 24.3. *Suppose $-A$ generates a bounded holomorphic strongly continuous semigroup of angle θ, and $n(\frac{\pi}{2} - \theta) < \frac{\pi}{2}$. Then $-A^n$ generates a bounded holomorphic strongly continuous semigroup of angle $\frac{\pi}{2} - n(\frac{\pi}{2} - \theta)$.*

Remark 24.4. Note that this is saying that, if A is of type θ (see 20.2(7)), then A^n is of type $n\theta$.

Theorem 24.5. *Suppose* $-A$ *generates an exponentially bounded holomorphic* k-*times integrated semigroup of angle* θ, $p(t) = t^n + q(t)$, *where* q *is a polynomial of degree less than* n, *and* $n(\frac{\pi}{2} - \theta) < \frac{\pi}{2}$. *Then*

(a) *the operator* $-p(A)$ *generates an exponentially bounded holomorphic* $(k+1)$-*times integrated semigroup of angle* $\frac{\pi}{2} - n(\frac{\pi}{2} - \theta)$;

(b) *if* $\mathcal{D}(p(A))$ *is dense, then* $-p(A)$ *generates an exponentially bounded holomorphic* k-*times integrated semigroup of angle* $\frac{\pi}{2} - n(\frac{\pi}{2} - \theta)$; *and*

(c) *if* $k = 0$, *then* $-p(A)$ *generates a strongly continuous holomorphic semigroup of angle* $\frac{\pi}{2} - n(\frac{\pi}{2} - \theta)$.

Corollary 24.6. *Suppose* p *is a polynomial with positive leading coefficient.*

(a) *If* $-A$ *generates a strongly continuous holomorphic semigroup of angle* $\frac{\pi}{2}$, *then* $-p(A)$ *generates a strongly continuous holomorphic semigroup of angle* $\frac{\pi}{2}$.

(b) *If* $-A$ *generates an exponentially bounded holomorphic* k-*times integrated semigroup of angle* $\frac{\pi}{2}$, *then* $-p(A)$ *generates an exponentially bounded holomorphic* $(k+1)$-*times integrated semigroup of angle* $\frac{\pi}{2}$.

The following theorem would follow from Theorem 24.5 and the fact that the square of the generator of a strongly continuous group generates a strongly continuous holomorphic semigroup, if $q(t)$ contained only even powers of t. Theorem 24.7 is more general, in that q may be any polynomial of degree less than $2n$.

Theorem 24.7. *Suppose* iA *generates a strongly continuous group, and* $p(t) = t^{2n} + q(t)$, *where* q *is a polynomial of degree less than* $2n$. *Then* $-p(A)$ *generates a strongly continuous holomorphic semigroup of angle* $\frac{\pi}{2}$.

The same argument, using Theorem 24.2(a) and (b), instead of (c), gives the following.

Theorem 24.8. *Suppose both* iA *and* $-iA$ *generate exponentially bounded* k-*times integrated semigroups and* p *is as in Theorem 24.7. Then*

(a) $-p(A)$ *generates an exponentially bounded holomorphic* $(k+1)$-*times integrated semigroup of angle* $\frac{\pi}{2}$.

(b) *If* $\mathcal{D}(p(A))$ *is dense, then* $-p(A)$ *generates an exponentially bounded holomorphic* k-*times integrated semigroup of angle* $\frac{\pi}{2}$.

159

Corollary 24.9. *Suppose p is an arbitrary polynomial.*

(a) *If both iA and $-iA$ generate exponentially bounded k-times integrated semigroups, then $-|p|^2(A)$ generates an exponentially bounded holomorphic $(k+1)$-times integrated semigroup of angle $\frac{\pi}{2}$.*

(b) *If, in addition to (a), $\mathcal{D}(|p|^2(A))$ is dense, then $-|p|^2(A)$ generates an exponentially bounded holomorphic k-times integrated semigroup of angle $\frac{\pi}{2}$.*

(c) *If iA generates a strongly continuous group, then $-|p|^2(A)$ generates a strongly continuous holomorphic semigroup of angle $\frac{\pi}{2}$.*

Example 24.10. Let $A \equiv i\frac{d}{dx}$, on $L^p(\mathbf{R})$ $(1 \leq p \leq \infty)$, with maximal domain, and $B \equiv (-1)^n(\frac{d}{dx})^{2n} + q(i\frac{d}{dx}) = A^{2n} + q(A)$, where q is a polynomial of degree less than $2n$.

(a) If $1 \leq p < \infty$, then B generates a strongly continuous holomorphic semigroup of angle $\frac{\pi}{2}$.

(b) If $p = \infty$, then B generates an exponentially bounded holomorphic once-integrated semigroup of angle $\frac{\pi}{2}$.

Example 24.11. Let $A \equiv \Delta$, the Laplacian, on $L^p(\mathbf{R}^n)(1 \leq p \leq \infty)$, with domain as in Example 20.2(2), or the Laplacian on $L^p(\Omega)$, where Ω is a bounded open set in $\mathbf{R}^n$ with smooth boundary, with $\mathcal{D}(A) \equiv W^{2,p}(\Omega) \cap W_0^{1,p}(\Omega)$. Let p be a polynomial with positive leading coefficient.

(a) If $1 \leq p < \infty$, then $-p(A)$ generates a strongly continuous holomorphic semigroup of angle $\frac{\pi}{2}$.

(b) If $p = \infty$, then $-p(A)$ generates an exponentially bounded holomorphic once-integrated semigroup of angle $\frac{\pi}{2}$.

Lemma 24.12. *Suppose $sp(A) \subseteq K$ and there exists finite M such that*

$$\|(w - A)^{-1}\| \leq M|w|^n, \qquad \forall w \notin K,$$

and p is an Nth degree polynomial such that $p(0) = 0$. Then $sp(p(A)) \subseteq p(K)$, with

$$\|(z - p(A))^{-1}\| \leq M^N|z|^n,$$

for all $z \notin p(K)$.

Proof: Let $V \equiv p(K)$.

Suppose z is not in V. Let $\{w_j\}_{j=1}^N$ be the (not necessarily distinct) zeroes of $z - p(w)$, that is,

$$z - p(w) = \prod_{j=1}^N (w_j - w), \qquad \forall\, \text{complex } w.$$

We have

$$z - p(A) = \prod_{j=1}^N (w_j - A).$$

For any j, since $p(w_j) = z$ is not in V, w_j is not in K. Thus, for $1 \leq j \leq N$, $(w_j - A)$ is invertible, and $\|(w_j - A)^{-1}\| \leq M|w_j|^{k-1}$. Thus, $z - p(A)$ is invertible, and we obtain the following upper bound for $(z - p(A))^{-1}$.

$$\|(z - p(A))^{-1}\| \leq \prod_{j=1}^N \|(w_j - A)^{-1}\|$$

$$\leq M^N \prod_{j=1}^N |w_j|^{k-1}$$

$$= M^N |z|^{k-1}.$$

■

Proof of Theorem 24.2: Parts (a) and (b) follow from Lemma 24.12 and Theorem 21.13. Part (c) follows from Lemma 24.12 and the fact that, when an operator B is densely defined, then $-B$ generates a bounded strongly continuous holomorphic semigroup if and only if there exists $\theta < \frac{\pi}{2}$ such that $sp(B) \subseteq \overline{S_\theta}$ and $\{\|z(z - B)^{-1}\| \mid z \notin \overline{S_\theta}\}$ is bounded. ■

Proof of Corollary 24.3: For this proof, let us write BHS for bounded strongly continuous holomorphic semigroup. Suppose $\frac{\pi}{2} > \psi > n(\frac{\pi}{2} - \theta)$. Then, since $-A$ generates a BHS of angle θ, the spectrum of A is contained in $S_{\frac{\psi}{n}}$, and $\{\|z(z - A)^{-1}\| \mid z \notin S_{\frac{\psi}{n}}\}$ is bounded.

Let $q(t) = t^n$. Since A generates a BHS, $\mathcal{D}(q(A))$ is dense. Also $q(S_{\frac{\psi}{n}})$ is contained in S_ψ, so by Theorem 24.2(c), $-A^n = -q(A)$ generates a BHS of angle $\frac{\pi}{2} - \psi$, whenever $\frac{\pi}{2} > \psi > n(\frac{\pi}{2} - \theta)$. This implies that $-A^n$ generates a BHS of angle $\frac{\pi}{2} - n(\frac{\pi}{2} - \theta)$. ■

In order to apply Theorem 24.2 to more general polynomials of other generators, we need some elementary lemmas.

Lemma 24.13. *Suppose E is a subset of the complex plane, and $\theta \geq 0$. Then*

$$\overline{\lim_{R \to \infty}} \sup\{|arg(z)| \mid z \in E, |z| = R\} \leq \theta$$

if and only if for all $\psi > \theta$, there exists real c_ψ such that E is contained in $c_\psi + S_\psi$.

Proof: Suppose the $\overline{\lim}$ inequality holds, and $\psi > \theta$. There exists finite M such that $|arg(z)| < \psi$, when z is in E and $|z| \geq M$. Thus, E is contained in $S_\psi \cup \{z \in C \mid |z| \leq M\}$, which may be shown to be contained in $-M(1 + \cot \psi) + S_\psi$.

Conversely, suppose that, for all $\psi > \theta$, there exists real c_ψ such that E is contained in $c_\psi + S_\psi$. For any $\psi \leq \pi$, it is not hard to see that

$$\overline{\lim_{R \to \infty}} \sup\{|arg\,(z)| \mid z \in c_\psi + S_\psi, |z| = R\}$$

equals ψ. Thus $\overline{\lim}_{R \to \infty} \sup\{|arg(z)| \mid z \in E, |z| = R\} \leq \psi$, for all $\psi > \theta$, which concludes the proof. ∎

Lemma 24.14. *If $p(t) = t^n + q(t)$, where q is a polynomial of degree less than n, $\theta \geq 0$ and c is real, then, for all $\psi > n\theta$, there exists real c_ψ such that $p(c + S_\theta)$ is contained in $c_\psi + S_\psi$.*

Proof: Clearly $\lim_{|z| \to \infty} \frac{p(c+z)}{z^n} = 1$. Thus,

$$\overline{\lim_{R \to \infty}} \sup\{|arg(z)| \mid z \in p(c + S_\theta), |z| = R\}$$
$$= \overline{\lim_{R \to \infty}} \sup\{|arg(p(c + z))| \mid z \in S_\theta, |z| = R\}$$
$$= \overline{\lim_{R \to \infty}} \sup\{|arg(z^n)| \mid z \in S_\theta, |z| = R\}$$
$$= n\theta.$$

Applying Lemma 24.13 now gives the result. ∎

Lemma 24.15. *Suppose K equals $-c + S_\theta \cup c - S_\theta$, where c and θ are nonnegative, and $p(t) = t^{2n} + q(t)$, where q is a polynomial of degree less than $2n$. Then, for all $\psi > 2n\theta$, there exists real c_ψ such that $p(K)$ is contained in $c_\psi + S_\psi$.*

Proof: Let $K^+ = -c + S_\theta$. Since $K = K^+ \cup -K^+$, it is sufficient, by Lemma 24.13, to show that $\overline{\lim}_{r \to \infty} \sup\{|arg(p(z))| \mid z \in \pm K^+\} = 2n$; this follows exactly as in the proof of Lemma 24.14. ∎

Proof of Theorem 24.5: Suppose $\frac{\pi}{2} > \psi > n(\frac{\pi}{2} - \theta)$. Choose ϕ such that $\frac{\psi}{n} > \phi > \frac{\pi}{2} - \theta$. Since $-A$ generates an exponentially bounded holomorphic k-times integrated semigroup of angle θ, there exists real c such that the spectrum of A is contained in $c + S_\phi$, and $\{\|w^{1-k}(w - A)^{-1}\| \mid w \notin c + S_\phi\}$ is bounded.

By Lemma 24.14, there exists real c_ψ such that $p(c + S_\phi)$ is contained in $c_\psi + S_\psi$. By Theorem 24.2(a), $c_\psi I - p(A)$ generates an exponentially bounded holomorphic $(k + 1)$-times integrated semigroup of angle $\frac{\pi}{2} - \psi$.

Thus, whenever $\frac{\pi}{2} > \psi > n(\frac{\pi}{2} - \theta)$, $-p(A)$ generates an exponentially bounded holomorphic $(k+1)$-times integrated semigroup of angle $\frac{\pi}{2} - \psi$. This implies (a).

The same argument, using Theorem 24.2(b), implies (b).

For (c), note that, since $-A$ generates a strongly continuous semigroup, $\mathcal{D}(p(A))$ is dense. Thus the argument above, with Theorem 24.2(c), implies (c). ∎

Proof of Theorem 24.7: Suppose $\frac{\pi}{2} > \phi > 0$. Since iA generates a strongly continuous group, there exists positive r such that the spectrum of A is contained in the horizontal strip $\{z \in \mathbf{C} \mid |\mathrm{Im}(z)| < r\}$, with $\{\|\mathrm{Im}(z)(z - A)^{-1}\| \mid |\mathrm{Im}(z)| \geq r\}$ bounded.

Let $c = r \cot(\frac{\phi}{4n})$, $K = (-c + S_{\frac{\phi}{4n}}) \cup (c - S_{\frac{\phi}{4n}})$. Since $\{z \in \mathbf{C} \mid |\mathrm{Im}(z)| < r\}$ is contained in K, and $\{|\frac{z}{\mathrm{Im}(z)}| \mid z \notin K\}$ is bounded, it follows that $\{\|z(z - A)^{-1}\| \mid z \notin K\}$ is bounded.

By Lemma 24.15, there exists real c_ϕ such that $p(K)$ is contained in $c_\phi + S_\phi$.

Since iA generates a strongly continuous group, $\mathcal{D}(p(A))$ is dense. By Theorem 24.2(c), $c_\phi I - p(A)$ generates a bounded strongly continuous holomorphic semigroup of angle $\frac{\pi}{2} - \phi$. Thus, for any positive ϕ, $-p(A)$ generates a strongly continuous holomorphic semigroup of angle $\frac{\pi}{2} - \phi$, so that $-p(A)$ generates a strongly continuous holomorphic semigroup of angle $\frac{\pi}{2}$. ∎

XXV. ITERATED ABSTRACT CAUCHY PROBLEMS

Perhaps the most unified treatment of higher order abstract Cauchy problems is via the *iterated abstract Cauchy problem*

$$(a) \quad \prod_{k=1}^{n} \left(\frac{d}{dt} - A_k \right) u(t) = 0 \quad (t \geq 0)$$

$$(b) \quad u^{(i-1)}(0) = x_i \quad (1 \leq i \leq n). \tag{25.1}$$

In this chapter, we show that (25.1) has a unique solution, for all x_i in a large set, whenever A_k generates a regularized semigroup, for $1 \leq k \leq n$. The empty product $\prod_{k=1}^{0} \left(\frac{d}{dt} - A_k \right)$ denotes the identity operator.

Corollary 25.7 guarantees solutions for a dense set of initial data. By a *solution* of (25.1) we mean (as in [S]) u such that $\prod_{k=1}^{j} (d/dt - A_k) u(t) \in C^1[0, \infty) \cap \mathcal{D}(A_{j+1})$, for $t \geq 0$, $0 \leq j < n$, satisfying (25.1).

Definition 25.2. We will say that (25.1) is *nicely solvable with respect to C* if there exists a unique solution, for all $x_i \in C^n(\mathcal{D})$, where

$$\mathcal{D} \equiv \bigcap \{ \mathcal{D}(A_{k_1} A_{k_2} \cdots A_{k_m}) \mid k_i \neq k_j \text{ when } i \neq j, 1 \leq k_i \leq n \}$$

such that $C^{-j} \left(\prod_{k=1}^{j} (d/dt - A_k) u_m(t) \right)$ converges to zero, as $m \to \infty$, uniformly on compact sets, for $0 \leq j < n$, whenever $u_m(0) \in C^n(\mathcal{D})$ and $\lim_{m \to \infty} \| C^{-(j+1)} \prod_{k=1}^{j} (d/dt - A_k) u_m(0) \| = 0$, for $0 \leq j < n$.

Theorem 25.3. *Suppose, for $1 \leq k \leq n$, that A_k generates a C-regularized semigroup. Then (25.1) is nicely solvable with respect to C.*

Under some additional hypotheses on $\{A_k\}_{k=1}^{n}$, we obtain a much simpler regularized semigroup version of continuous dependence on the initial data, while representing the solutions explicitly, in a "regularized semigroup d'Alembert form,"

$$u(t) = \sum_{k=1}^{n} W_k(t) y_k,$$

where $\{W_k(t)\}_{t \geq 0}$ is the regularized semigroup generated by A_k, generalizing [Go-S4].

164

Theorem 25.4. *Suppose*

(1) *for $1 \leq k \leq n$, A_k generates a C-regularized semigroup, $\{W_k(t)\}_{t\geq 0}$;*

(2) *for $1 \leq i < l \leq n, (A_i - A_l)$ is injective, on $\mathcal{D}(A_i) \cap \mathcal{D}(A_l)$; and*

(3) *$W_i(s)W_l(t) = W_l(t)W_i(s)$, for $s, t \geq 0, 1 \leq i, l \leq n$.*

Then, if u is a solution of (25.1), with

$$x_i \in C\left(\prod_{1\leq i<l\leq n}(A_i - A_l)^n \left(\mathcal{D}\left(\prod_{j=1}^n A_j^{n^2+1}\right)\right)\right),$$

for all i, then there exists $\{y_j\}_{j=1}^n$ such that

$$u(t) = \sum_{j=1}^n W_j(t)y_j.$$

For our next theorem, we will use the following terminology.

Definition 25.5. Let $\mathcal{D}_n \equiv \bigcap \{\mathcal{D}\left(A_k^{n-i}\prod_{l=1}^i A_{k_l}\right) \mid 1 \leq i, k, k_l \leq n\}$,

$$\mathcal{A}_n \equiv \begin{bmatrix} 1 & \cdots & 1 \\ A_1 & \cdots & A_n \\ \vdots & & \\ A_1^{n-1} & \cdots & A_n^{n-1} \end{bmatrix}, \quad \mathcal{D}(\mathcal{A}_n) \equiv \times_{k=1}^n \mathcal{D}(A_k^{n-1})$$

$$A \equiv \begin{bmatrix} A_1 & & 0 \\ & \ddots & \\ 0 & & A_n \end{bmatrix}, \quad \|z\| \equiv \sum_{k=1}^n \|z_k\|.$$

Note that $\mathcal{D}_n$ is the set in which $y_1, \ldots, y_n$ must lie, to guarantee that u, when written in the regularized semigroup d'Alembert form above, is a solution of any permutation

$$\prod_{k=1}^n \left(\frac{d}{dt} - A_{\sigma(k)}\right)u = 0,$$

where σ is a permutation of $\{1, \ldots, n\}$, of (25.1). The operator $\mathcal{A}_n$ performs the change of variables, $\vec{x} = \mathcal{A}_n C\vec{y}$, from the parameters $y_1, \ldots, y_n$ to the initial data.

165

Theorem 25.6. *Suppose $\{A_k\}_{k=1}^n, \{W_k(t)\}_{k=1}^n$, are as in Theorem 25.4, and $\cap_{k=1}^n \rho_C(A_k)$ is nonempty. Then, for all $\vec{x} \in C\mathcal{A}_n(\mathcal{D}_n^n)$, there exists a unique solution of (25.1), given by*

$$u(t) = \sum_{k=1}^n W_k(t) y_k,$$

where $\vec{y} = (C\mathcal{A}_n)^{-1}\vec{x}$. There exists finite M such that

$$\|u^{(l)}(t)\| \le M \left(\sup\{\|W_k(t)\| \mid 1 \le k \le n\} \|A^l (C\mathcal{A}_n)^{-1}\vec{x}\| \right),$$

whenever $0 \le l \le n, t \ge 0$.

When $\mathcal{D}(A_k)$ and $Im(A_i - A_l)$ are dense, it may be shown that $Im(\mathcal{A}_n)$ is dense. Thus, since the generator of a C-regularized semigroup is densely defined whenever $Im(C)$ is dense (see Theorem 3.4), we have the following.

Corollary 25.7. *Suppose $\{A_k\}_{k=1}^n$ and $\{W_k(t)\}_{k=1}^n$ are as in Theorem 25.6 and $Im(C)$ and $Im(A_i - A_l)$ are dense, whenever $1 \le i < l \le n$. Then there exists $\vec{\mathcal{D}} \equiv C\mathcal{A}_n(\mathcal{D}_n^n)$, a dense subspace of X^n, such that for all $\vec{x} \in \vec{\mathcal{D}}$, there exists a unique solution of (25.1), given by*

$$u(t) = \sum_{k=1}^n W_k(t) y_k,$$

where $\vec{y} = (C\mathcal{A}_n)^{-1}\vec{x}$.

Example 25.8. HIGHER ORDER ABSTRACT CAUCHY PROBLEMS. We consider

$$\begin{aligned} w^{(n)}(t) &= A(w(t)) \quad (t \ge 0), \\ w^{(i-1)}(0) &= x_i \quad (1 \le i \le n). \end{aligned} \tag{25.9}$$

By a *solution* we mean n-times continuously differentiable $w : [0, \infty) \to X$ such that $w^{(i)}(t) \in \mathcal{D}(A)$, for $0 \le i < n$, satisfying (25.9).

It is well known that, for $n > 2$, (25.9) fails to be well-posed, in the sense of strongly continuous semigroups, unless A is bounded (see [Hil], [F2]). However, Theorem 8.2, along with Theorem 25.6 and the factorization

$$(w^{(n)} - A) = \prod_{k=1}^n \left(\frac{d}{dt} - e^{\frac{2\pi k i}{n}} A^{\frac{1}{n}} \right) w, \quad w^{(i-1)}(0) = x_i, \tag{25.10}$$

soon yields Theorem 25.11 below.

Theorem 25.11. *Suppose λA has dense range and generates a bounded strongly continuous semigroup, for some $\lambda \in \mathbf{C}$, and $n > 1$. Then there exists $\vec{D}$, a dense subset of X^n, such that, for all $\vec{x} \in \vec{D}$, there exists a solution of (25.9), that extends to an entire function, given by*

$$w(t) = \sum_{k=1}^{n} W((-\frac{1}{\lambda})^{\frac{1}{n}} e^{\frac{2\pi k i}{n}} t) y_k,$$

where $1 < \alpha < 2$, $\{W(t)\}_{t \geq 0}$ is an entire $e^{-(-\lambda A)^{\frac{\alpha}{n}}}$-regularized group, generated by $(-\lambda A)^{\frac{1}{n}}$.

Equation (25.10) is nicely solvable with respect to $e^{-(-\lambda A)^{\alpha/n}}$. There exists $M < \infty$ such that

$$\|e^{-(-\lambda A)^{\alpha/n}} w^{(l)}(t)\| \leq M \sup_{|z|=t} \|W(z)\| \sum_{i=1}^{n} \sup_{0 \leq k \leq n-1} \|(-\lambda A)^{\frac{l-k}{n}} x_i\|,$$

for $0 \leq l \leq n, t \geq 0, \vec{x} \in \vec{D}$.

See also Chapter XIV.

Example 25.12. THE CAUCHY PROBLEM FOR THE LAPLACE EQUATION. As a special case of Theorem 25.11, we are able to treat the Cauchy problem for the Laplace equation, on a cylinder $[0, \infty) \times D$, with Dirichlet boundary conditions,

$$(\frac{d}{dx})^2 u(x, \vec{y}) + \Delta u(x, \vec{y}) = 0 \quad (\vec{y} \in D, \ x > 0)$$

$$u(x, \vec{y}) = 0 \quad (\vec{y} \in \partial D, \ x > 0)$$

$$u(0, \vec{y}) = f_1(\vec{y}) \quad (\vec{y} \in D) \tag{25.13}$$

$$\frac{\partial u}{\partial x}(0, \vec{y}) = f_2(\vec{y}),$$

where D is a bounded open set in $\mathbf{R}^n$ with smooth boundary ∂D, and on the strip $[0, \infty) \times [0, 1]$, with periodic boundary conditions,

$$\left(\frac{d}{dx}\right)^2 u(x, y) + \left(\frac{d}{dy}\right)^2 u(x, y) = 0 \quad (0 \leq y \leq 1, \ x > 0)$$

$$u(0, y) = f_1(y), \quad \frac{\partial u}{\partial x}(0, y) = f_2(y), \tag{25.14}$$

$$u(x, 0) = \alpha u(x, 1),$$

where $|\alpha| = 1$.

The following theorem is a consequence of Theorem 25.11, by choosing A equal to $-B^2$, where $B \equiv \frac{d}{dy}$, on $Y \equiv L^p[0,1]$ $(1 \le p < \infty)$, $\mathcal{D}(B) = \{f \in Y \mid f' \in Y, \ f(0) = \alpha f(1)\}$ $(|\alpha| = 1)$, for (a), and, for (b), by choosing A equal to $-\triangle$ on $L^p(D), \mathcal{D}(A) \equiv W^{2,p}(D) \cap W_0^{1,p}(D)$.

Theorem 25.15. *Suppose $1 \le p < \infty$.*

(a) *If $X = L^p[0,1]$, then (25.14) has a unique solution, for all f_1, f_2 in a dense subspace of X, that extends to an entire function.*

(b) *If $X = L^p(D)$, then (25.13) has a unique solution, for all f_1, f_2 in a dense subspace of X, that extends to an entire function.*

See also Chapter IX.

Example 25.16. WEIGHTED COMPOSITION GROUPS. We may also apply our results to

$$(\frac{d}{dt})^2 u(t,x) = (\frac{d}{dx})^2 u(t,x) + \frac{\alpha}{x}\frac{d}{dx}u(t,x) + \frac{\alpha^2 - \alpha}{x^2}u(t,x)\,(t,x \in \mathbf{R})$$

$$u(0,x) = f_1(x), \ \frac{d}{dt}u(0,x) = f_2(x)\,(x \in \mathbf{R}),$$

$$\text{(25.17)}$$

where $|\alpha| < 1, \alpha \ne 0$, as follows.

We want to consider an operator $G \equiv \frac{d}{dx} + \frac{\alpha}{x}$, on an appropriate subspace of $L^1(\mathbf{R})$. Formally, G generates the following group (see [Go2, 8.14]):

$$(e^{tG}f)(x) \equiv (\frac{x+t}{x})^\alpha f(x+t).$$

It is not hard to see that e^{tG} does not map $L^1(\mathbf{R})$ into itself. However, if we integrate once,

$$(S(t)f)(x) \equiv \int_0^t (e^{sG}f)(x)\,ds,$$

then elementary (but long) calculations show that, for each $t \in \mathbf{R}$, $S(t)$ maps $L^1(\mathbf{R})$ into $L^1(\mathbf{R})$, thus, since it is a positive operator, is bounded. A similar calculation, with dominated convergence, shows that $S(t)$ is strongly continuous.

We define G to be the generator of the once-integrated group $\{S(t)\}_{t\in\mathbf{R}}$; this is equivalent to being the generator of a $(\lambda - G)^{-1}$-regularized group (see Chapter XVIII).

168

Thus we may apply Theorem 25.11 by writing (25.17) as

$$(\frac{d}{dt})^2 u(t,\cdot) = G^2 u(t,\cdot),$$

to obtain the following.

Theorem 25.18. *Suppose $X \equiv L^1(\mathbf{R})$. Then (25.17) has a unique solution, for all f_1, f_2 in a dense subspace of X, that extends to an entire function.*

Remark 25.19. The operators $\frac{d}{dx}$ and $\frac{\alpha}{x}$ scale the same way; that is, if $\lambda > 0$ and U_λ is the unitary transformation on $L^2(\mathbf{R})$

$$(U_\lambda u)(x) \equiv \sqrt{\lambda} u(\lambda x),$$

then $M(U_\lambda u) = \lambda U_\lambda(Mu)$, for both $M = \frac{d}{dx}$ and $M = \frac{\alpha}{x}$. The fact that Δ and $\frac{c}{|x|^2}$ scale the same is very important in quantum mechanics; see [Bar-Go] and [Go-Sv].

Operators similar to G above appear in [A1, section 3].

To prove Theorem 25.3, we generalize the construction in [S]. On X^n, define A to be a diagonal operator matrix,

$$A \equiv \begin{bmatrix} A_1 & & 0 \\ & \ddots & \\ 0 & & A_n \end{bmatrix},$$

$$\mathcal{D}(A) \equiv \bigotimes_{k=1}^n \mathcal{D}(A_k),$$

$$B \equiv \begin{bmatrix} 0 & C & & 0 \\ & \ddots & \ddots & \\ & & & C \\ 0 & & & 0 \end{bmatrix},$$

the operator matrix with C in the super diagonal, zeroes elsewhere.

Consider

$$\frac{d\vec{u}}{dt} = (A+B)\vec{u}(t) \quad (t \geq 0), \tag{25.20}$$

where $\vec{u} \equiv (u_1,\ldots,u_n)$.

169

Lemma 25.21. *The following are equivalent.*

(a) $\vec{u}$ *is a solution of* (25.20).

(b) u_1 *is a solution of* (25.1)(a), *with*

$$\prod_{k=1}^{j} \left(\frac{d}{dt} - A_k \right) u_1(t) \in C^j \left(\mathcal{D}(A_{j+1}) \right),$$

$$u_{j+1}(t) = C^{-j} \left(\prod_{k=1}^{j} \left(\frac{d}{dt} - A_k \right) u_1(t) \right),$$

for $t \geq 0, 0 \leq j < n$.

Proof: $\vec{u}$ is a solution of (25.20) if and only if

$$\frac{d}{dt} u_j(t) = A_j u_j(t) + C u_{j+1}(t),$$

for $1 \leq j < n$, and

$$\frac{d}{dt} u_n(t) = A_n u_n(t).$$

This is equivalent to

$$C^j u_{j+1} = \prod_{k=1}^{j} \left(\frac{d}{dt} - A_k \right) u_1, \quad 0 \leq j < n,$$

and

$$0 = \prod_{k=1}^{n} \left(\frac{d}{dt} - A_k \right) u_1. \quad \blacksquare$$

Mimicking the construction of the perturbed semigroup, when an operator that generates a strongly continuous semigroup is perturbed by a bounded operator, gives us the following (see [dL9, Theorem 6.1]).

Lemma 25.22. *Suppose A generates a C-regularized semigroup, $BC = CB$ and $C^{-1}B \in B(X)$. Then $(A + B)$ generates a C-regularized semigroup.*

Proof of Theorem 25.3: To prove existence, suppose $x_1, \ldots, x_n$ are in $C^n(\mathcal{D})$.

For each j, there exists $\{p_{i,j}\}_{i=0}^{j}$, polynomials in j noncommuting variables, such that

$$\prod_{k=1}^{j}\left(\frac{d}{dt}-A_k\right) = \sum_{i=0}^{j} p_{i,j}(A_1,\cdots,A_j)\left(\frac{d}{dt}\right)^i. \qquad (25.23)$$

Since each $p_{i,j}$ is a sum of products $A_{k_1}\dots A_{k_m}$, where $k_i \neq k_\ell$ when $i \neq \ell$, and $1 \leq k_i \leq j$, we have

$$p_{i,j}(A_1,\dots,A_j)x_\ell \in C^n\left(\mathcal{D}(A_{j+1})\right), \qquad (25.24)$$

for $0 \leq j < n$, $1 \leq \ell \leq n$.

Thus we may make the following definition. Define $\vec{y}$, in X^n, by $\vec{y} \equiv (y_1,\dots,y_n)$, where $y_1 \equiv x_1$, and

$$y_{j+1} \equiv C^{-j}\left(\sum_{i=0}^{j} p_{i,j}(A_1,\dots,A_j)x_{i+1}\right), \qquad (25.25)$$

for $1 \leq j < n$.

Note that, by (25.24), $\vec{y} \in C(\mathcal{D}(A))$.

By Lemma 25.21, $(A+B)$ generates a C-regularized semigroup on X^n. Thus, by Theorem 3.5, there exists a unique solution, $\vec{u}$, of (25.20), such that $\vec{u}(0) = \vec{y}$. By Lemma 25.21, u_1 is a solution of (25.1)(a). By Lemma 25.21, (25.23) and (25.25),

$$\sum_{i=0}^{j} p_{i,j}(A_1,\dots,A_j)u_1^{(i)}(0) = \sum_{i=0}^{j} p_{i,j}(A_1,\dots,A_j)x_{i+1},$$

for $0 \leq j < n$.

Since $p_{j,j}(A_1,\dots,A_j) = I$, for all j, this enables us to solve, inductively, for $u_1^{(i)}(0)$, to obtain $u_1^{(i)}(0) = x_{i+1}$, for $0 \leq i < n$, so that u_1 is a solution of (25.1).

To establish uniqueness, suppose u is a solution of (25.1), with $x_i = 0$, for $1 \leq i \leq n$. Let

$$v_1(t) \equiv C^{2n}(u(t)),\, v_{j+1}(t) \equiv C^{2n-j}\left(\prod_{k=1}^{j}\left(\frac{d}{dt}-A_k\right)u(t)\right),$$

for $1 \leq j < n$.

By Lemma 25.21, $\vec{v} \equiv (v_1,\dots,v_n)$ is a solution of (25.20). Since $(A+B)$ generates a C-regularized semigroup, and $\vec{v}(0) = 0$, it follows from

171

Theorem 3.5 that $\bar{v}(t) = 0$, for all $t \geq 0$, so that $C^{2n}(u(t)) = 0$, for all $t \geq 0$. Since C is injective, $u(t) \equiv 0$, as desired.

To see that (25.1) is nicely solvable with respect to C, suppose $\{u_m\}_{m=1}^{\infty}$ is a sequence of solutions such that, for $0 \leq j < n$,

$$\lim_{m \to \infty} \left\| C^{-(j+1)} \prod_{k=1}^{j} \left(\frac{d}{dt} - A_k \right) u_m(0) \right\| = 0.$$

For any m, let $\vec{u}_m \equiv ((u_m)_1, \ldots, (u_m)_n)$, where $(u_m)_1 \equiv u_m$, $(u_m)_{j+1} \equiv C^{-j} \left(\prod_{k=1}^{j} \left(\frac{d}{dt} - A_k \right) u_m \right)$ $(0 \leq j < n)$. By Lemma 25.21 and Theorem 3.13, since $\lim_{m \to \infty} \| C^{-1} \vec{u}_m(0) \| = 0$, it follows that $\vec{u}_m(t)$ converges to zero, uniformly on compact sets, so that, by looking at the components of $\vec{u}_m$, we obtain the desired convergence. ∎

Lemma 25.26. *Suppose $\{W(t)\}_{t \geq 0}$ is a C-regularized semigroup generated by A, $f : [0, \infty) \to Im(C)$ has the property that $t \mapsto C^{-1} f(t)$ is continuously differentiable, and $x \in C(\mathcal{D}(A))$. Then*

$$u'(t) = A(u(t)) + f(t) \, (t \geq 0), \, u(0) = x \qquad (*)$$

has the unique solution

$$u(t) = W(t)C^{-1}x + \int_0^t W(t-s)C^{-1}f(s) \, ds. \qquad (**)$$

Proof: Suppose u is given by (**). By Theorem 3.4, for all $t > 0$, $\int_0^t W(t-s)C^{-1}f(s) \, ds \in \mathcal{D}(A)$, with

$$A\left(\int_0^t W(t-s)C^{-1}f(s) \, ds \right) = W(t)C^{-1}f(0) - f(t)$$

$$+ \int_0^t W(t-s)C^{-1}f'(s) \, ds,$$

so that

$$A(u(t)) = W'(t)C^{-1}x + W(t)C^{-1}f(0) - f(t) + \int_0^t W(t-s)C^{-1}f'(s) \, ds,$$

which a calculation shows to be equal to $u'(t) - f(t)$.

For uniqueness, suppose u satisfies (*). Then, for $0 \leq s \leq t$, $\frac{d}{ds}W(t-s)u(s) = W(t-s)f(s)$, so that $W(0)u(t) - W(t)u(0) = \int_0^t W(t-s)f(s) \, ds$, or

$$Cu(t) = C\left[W(t)C^{-1}x + \int_0^t W(t-s)C^{-1}f(s) \, ds \right],$$

so that, since C is injective, u is as in (**). ∎

Lemma 25.27. *Suppose, for $i = 1, 2$, that A_i generates a C-regularized semigroup $\{W_i(t)\}_{t \geq 0}$ and $W_1(s)W_2(t) = W_2(t)W_1(s)$, for all $s, t \geq 0$. Then, for all $t > 0, x \in \mathcal{D}(A_1) \cap \mathcal{D}(A_2)$,*

$$\int_0^t W_1(t-s)W_2(s)(A_2 - A_1)x = C\left(W_2(t) - W_1(t)\right)x.$$

Proof: We calculate:

$$\int_0^t W_1(t-s)W_2(s)(A_2 - A_1)x \, ds = \int_0^t \frac{d}{ds}\left(W_1(t-s)W_2(s)x\right) ds$$

$$= W_1(0)W_2(t)x - W_1(t)W_2(0)x$$

$$= C(W_2(t) - W_1(t))x.$$

∎

Proof of Theorem 25.4: For $0 \leq j \leq n$, let

$$v_j(t) \equiv \prod_{k=1}^{j}\left(\frac{d}{dt} - A_k\right) u(t).$$

We will show, by induction on k, that, for $0 \leq k \leq n$, there exists

$$\{y_{k,j}\}_{j=0}^{k-1} \in \bigcup\{ \prod_{1 \leq i < l \leq n}(A_i - A_l)^{m_{i,l}}\left(\mathcal{D}\left(\prod_{j=1}^{n}A_j^{n^2}\right)\right) \mid n - k < m_{i,l} \leq n\}$$

such that

$$v_{n-k}(t) = \sum_{j=0}^{k-1} W_{n-j}(t)y_{k,j}. \qquad (*)$$

For $k = 0$, (*) is clear, since $v_n = 0$.
Suppose now that (*) holds for a fixed k. Then

$$(\frac{d}{dt} - A_{n-k})v_{n-(k+1)}(t) = \sum_{j=0}^{k-1} W_{n-j}(t)y_{k,j},$$

so by Lemma 25.26,

$$Cv_{n-(k+1)}(t) = W_{n-k}(t)v_{n-(k+1)}(0) + \int_0^t W_{n-k}(t-s)\sum_{j=0}^{k-1} W_{n-j}(s)y_{k,j}$$

$$= W_{n-k}(t)v_{n-(k+1)}(0) + C\sum_{j=0}^{k-1}[W_{n-j}(t) - W_{n-k}(t)]\left(A_{n-j} - A_{n-k}\right)^{-1}y_{k,j},$$

by Lemma 25.27, completing the induction, since C is injective and

$$v_{n-(k+1)}(0) \in C \left(\prod_{1 \le i < l \le n} (A_i - A_l)^n \left(\mathcal{D} \left(\prod_{j=1}^n A_j^{n^2} \right) \right) \right).$$

Since $u(t) = v_0(t)$, this completes the proof. ∎

Proof of Theorem 25.6: It's clear that $u(t)$ is a solution of (25.1). For uniqueness, suppose $v(t)$ is a solution of (25.1), with $\vec{x} = \vec{0}$. For $r \in \cap_{k=1}^n \rho_C(A_k)$, let

$$R \equiv \left[\prod_{1 \le i < l \le n} (A_i - A_l)^n \right] \left[\prod_{k=1}^n \left((r - A_k)^{-1} C \right)^{n^2+1} \right], \quad w(t) \equiv R(v(t)).$$

By Theorem 25.4, there exists $\{y_j\}_{j=1}^n$ such that $w(t) = \sum_{j=1}^n W_j(t) y_j$. Thus $\vec{0} = \mathcal{A} \left(\prod_{k=1}^n \left((r - A_k)^{-1} C \right)^{n-1} \vec{y} \right)$. Hypothesis (2) of Theorem 25.4 implies that $\mathcal{A}$ is injective, thus $\vec{y}$, and hence $w(t)$, equals 0, for all $t \ge 0$, since $(r - A_k)^{-1} C$ is injective. Since R is injective, $v(t) = 0$, for all $t \ge 0$, as desired. ∎

Proof of Theorem 25.11: Since λA has dense range and generates a bounded strongly continuous semigroup, it is injective.

Let $(-\lambda A)^{\frac{1}{n}}$ be the unique n^{th} root of $(-\lambda A)$ such that $-(-\lambda A)^{\frac{1}{n}}$ generates a bounded strongly continuous holomorphic semigroup of angle $\frac{\pi}{2}(1 - \frac{1}{n})$. Let $A_k \equiv e^{\frac{2\pi k i}{n}} A^{\frac{1}{n}} \equiv (-\frac{1}{\lambda})^{\frac{1}{n}} e^{\frac{2\pi k i}{n}} (-\lambda A)^{\frac{1}{n}}$. By Theorem 8.2, $(-\lambda A)^{\frac{1}{n}}$ generates an entire $e^{-(-\lambda A)^{\frac{\alpha}{n}}}$-regularized group $\{W(z)\}_{z \in \mathbb{C}}$; thus for each k, A_k generates an $e^{-(-\lambda A)^{\frac{\alpha}{n}}}$-regularized semigroup, $\{W((-\frac{1}{\lambda})^{\frac{1}{n}} e^{\frac{2\pi k i}{n}} t)\}_{t \ge 0}$.

The factorization (25.10) and Theorem 25.3 now imply the nice solvability of (25.10).

To show the existence of the dense set, $\vec{\mathcal{D}}$, we may apply Corollary 25.7, once we verify that $Im(e^{-(-\lambda A)^{\alpha/n}})$ is dense. If $\alpha < \beta < \frac{\pi}{2\theta}$, where $\frac{\pi}{2} - \theta$ is the angle of the holomorphic semigroup generated by $-(-\lambda A)^{\frac{1}{n}}$, then it may be shown, with the usual Cauchy integral formula construction of fractional powers, that for all $\epsilon > 0$, $Im(e^{-\epsilon(-\lambda A)^{\frac{\beta}{n}}}) \subseteq Im(e^{-(-\lambda A)^{\alpha/n}})$. Since $\{e^{-\epsilon(-\lambda A)^{\frac{\beta}{n}}}\}_{\epsilon \ge 0}$ is a strongly continuous semigroup, this implies that $Im(e^{-(-\lambda A)^{\alpha/n}})$ is dense.

The d'Alembert form of the solution follows from Corollary 25.7. The estimates of $\|w^{(l)}(t)\|$ follow from Theorem 25.6 and the fact that each entry of $(\mathcal{A}_n)^{-1}$ has the form $c_{i,j}(A^{-\frac{1}{n}})^{k_{i,j}}$, where $c_{i,j}$ is a complex number, and $k_{i,j}$ is an integer between 0 and $n - 1$. ∎

174

XXVI. EQUIPARTITION OF ENERGY

In this chapter, we consider

$$w''(t) = B^2(w(t)) \quad (t \geq 0),$$
$$w(0) = x_1, \quad w'(0) = x_2. \tag{26.1}$$

Definition 26.2. As in [19], we will say that a solution, u, of (26.1), *admits sharp equipartition of energy* if

$$\lim_{t \to \infty} \frac{K(t)}{P(t)} = 1,$$

where $K(t)$, the kinetic energy, is defined to be $\|u'(t)\|^2$ and $P(t)$, the potential energy, is defined to be $\|Bu(t)\|^2$.

Remarks 26.3. Let $u = u(t,x)$ (respectively, $v = v(t,x)$) be the position (respectively, velocity) vector describing a wave in $\mathbf{R}^3$. Under some natural assumptions, u and v both satisfy the wave equation

$$w_{tt} = \Delta w$$

for $t \in \mathbf{R}, x \in \mathbf{R}^3$. If $\| \cdot \|$ denotes the norm in $L^2(\mathbf{R}^3)$, then

$$K(t) = \|w_t(t, \cdot)\|^2, \, P(t) = \|\nabla_x w(t, \cdot)\|^2$$

are the kinetic and potential energies (respectively the potential and kinetic energies) at time t for $w = u$ (respectively $w = v$). In Maxwell's equations in a vacuum, the kinetic energy of the electric vector is the potential energy of the magnetic vector and vice versa, cf. [Go-S1, p. 404].

Thus "kinetic" and "potential" energies are nice suggestive terms, but even in concrete physical contexts one should be careful in using them. Thus our use of these terms in the nonphysical context of this section should not be taken to have physical implications.

Theorem 26.4. *Suppose $Im(C)$ is dense, B is injective, has dense range and generates a C-regularized group $\{W(t)\}_{t\in\mathbf{R}}$ such that $\lim_{t\to\infty} W(t)x = 0$, for all $x \in X$. Then there exists a dense subspace, $\vec{D}$, of X^2, such that for all $\vec{x} \in \vec{D}$, there exists a unique solution of (26.1) that admits sharp equipartition of energy.*

175

Corollary 26.5. *Suppose B has dense range and generates a bounded strongly continuous holomorphic semigroup. Then there exists a dense subspace, $\vec{\mathcal{D}}$, of X^2, such that for all $\vec{x} \in \vec{\mathcal{D}}$, there exists a unique solution of (26.1) that admits sharp equipartition of energy.*

Example 26.6. The Cauchy problem for the Laplace equation admits sharp equipartition of energy. More precisely, for $1 \le p < \infty$, D as in (25.13), there exists a dense subspace, $\vec{\mathcal{D}}$, of $(L^p(D))^2$, such that for all $(f_1, f_2) \in \vec{\mathcal{D}}$, there exists a unique solution of (25.13) that admits sharp equipartition of energy.

This is a consequence of Corollary 26.5, by choosing B to be a square root of $-\Delta$, on $L^p(D)$, that generates a bounded strongly continuous holomorphic semigroup. Since Δ has dense range, B does also.

More generally, since the holomorphic semigroup generated by B is of angle $\frac{\pi}{2}$, we could replace Δ by $\lambda\Delta$, in (25.13), for any complex λ whose argument is less than π.

All these assertions are also true for (25.14).

Remark 26.7. It is interesting that in 26.4, 26.5 and 26.6, both $K(t)$ and $P(t)$ may go to infinity, as t goes to infinity, although their ratio converges to one.

Consider, for example, $Bf(s) \equiv sf(s)$, on $X \equiv C_0[0, \infty)$, with maximal domain. Then $w(t)(s) = e^{ts}y_1(s) + e^{-ts}y_2(s)$, for some $y_1, y_2 \in X$, so that $K(t)$ and $P(t)$ are greater than or equal to $\sup_{s \ge 0} |se^{ts}y_1(s)| - (te)^{-1}\|y_2\|_\infty$.

Proof of Theorem 26.4: First, note that, for any $x \in X$, $\{\|W(-t)x\|\}_{t \ge 0}$ is bounded below. This may be seen by supposing, for the sake of contradiction, that there exists a sequence of real numbers $\{t_k\}, x \in X$, such that $t_k \to \infty$ and $\|W(-t_k)x\| \to 0$. Then, since $C^2 x = W(t_k)W(-t_k)x$ and C is injective, this would imply that $\|W(t_k)\| \to \infty$, which contradicts the stability of $\{W(t)\}_{t \ge 0}$.

As in the proof of Theorem 25.11, we factor (26.1),

$$w''(t) - B^2(w(t)) = (\frac{d}{dt} - B)(\frac{d}{dt} + B)w(t),$$

and apply Corollary 25.7, to obtain a dense subspace $\vec{\mathcal{D}}$ such that, for all $\vec{x} \in \vec{\mathcal{D}}$, there exists a solution of (26.1) given by

$$u(t) = W(t)y_1 + W(-t)y_2,$$

for some $y_i \in X$. Thus $u'(t) = W(t)By_1 - W(-t)By_2$, and $Bu(t) = W(t)By_1 + W(-t)By_2$, so that, since $\|W(t)By_1\| \to 0$ and $\|W(-t)By_2\|$ is bounded below, as $t \to \infty$, u admits sharp equipartition of energy. $\blacksquare$

176

Proof of Corollary 26.5: By Theorem 8.2, there exists $\alpha > 1$ such that B generates an $e^{-(-B)^{\alpha}}$-regularized group. By Lemma 8.8, the image of $e^{-(-B)^{\alpha}}$ is dense. Since B generates a bounded strongly continuous semigroup, and has dense range, B is injective. Thus we may apply Theorem 26.4. ∎

XXVII. SIMULTANEOUS SOLUTION SPACE

Suppose $\{A_\alpha\}_{\alpha \in I}$ is a family of closed operators on a Banach space. As we did in Chapters IV and V with a single operator, we wish to construct maximal continuously embedded subspaces on which, for all α, A_α generates a strongly continuous semigroup.

This will be the *simultaneous solution space* for $\{A_\alpha\}_{\alpha \in I}$. In this chapter, we will consider the case where the simultaneous solution space is a locally convex space; as in Chapter IV, this space has both an operator-theoretic maximality (see Theorem 27.5) and a "pointwise" maximality (see Remark 27.4). In the next chapter, we will, as in Chapter V, consider the case where the simultaneous solution space is a Banach space.

Terminology 27.1. Throughout this chapter and the next two chapters, $\mathcal{A}$ will be a collection of closed operators $\{A_\alpha\}_{\alpha \in I}$. Since our operators will not commute, in general, we must specify the order of a product of operators:

$$\prod_{k=1}^{n} G_k \equiv G_1 G_2 \cdots G_n.$$

The empty product of operators $\prod_{k=1}^{0} G_k$ will mean the identity operator.

For a single closed operator, we introduced, in Definition 4.6, its *solution space*, and gave it a Frechet space topology, with respect to which the operator generated a strongly continuous locally equicontinuous semigroup (Theorem 4.8).

Let us write $Z(A)$ for the solution space of the closed operator A.

When the solutions of the abstract Cauchy problem (0.1) are unique, that is, there are no nontrivial solutions when $x = 0$, we may then define, for $t \geq 0$, the operator e^{tA} by

$$e^{tA} x \equiv u(t,x), \, \mathcal{D}(e^{tA}) \equiv Z(A).$$

There is no reason to believe that e^{tA} will be closed, or even closable, in general.

Definition 27.2. We will write $Z(\mathcal{A})$ for the *simultaneous solution space for $\mathcal{A}$*, which we define to be the set of all x in

$$\mathcal{D}(\mathcal{A}) \equiv \bigcap \mathcal{D}(\prod_{k=1}^{n} e^{t_k A_{\alpha_k}}),$$

where the intersection is taken over all finite sequences $\vec{t} \equiv < t_k >_{k=1}^n$, $\vec{\alpha} \equiv < \alpha_k >_{k=1}^n$, satisfying the following:

(1)

$$\|x\|_{A,\vec{a},\vec{b},\vec{\alpha}} \equiv \sup\{\|(\prod_{k=1}^n e^{t_k A_{\alpha_k}})x\| \mid t_k \in [a_k, b_k]\}$$

is finite, for all finite sequences of nonnegative real numbers $\vec{a}, \vec{b}$ and all finite sequences $\vec{\alpha}$, from I; and

(2) the map $t \mapsto e^{tA_\alpha}x$, from $[0, \infty)$ into $(\mathcal{D}(A), \| \ \|_{A,\vec{a},\vec{b},\vec{\alpha}})$ is continuous, for all $\alpha \in I, \vec{a}, \vec{b}, \vec{\alpha}$.

We give $Z(A)$ the locally convex topology generated by the seminorms $\| \ \|_{A,\vec{a},\vec{b},\vec{\alpha}}$.

Remarks 27.3. For a single operator, that is, $\mathcal{A} = \{A\}$, for some closed operator A, both (1) and (2) of Definition 27.2 are automatically satisfied. Another natural choice of seminorms would be

$$\|x\|_{\vec{t},\vec{\alpha}} \equiv \|(\prod_{k=1}^n e^{t_k A_{\alpha_k}})x\|,$$

for $t_k \geq 0, \alpha_k \in I$. Completeness of the space and local equicontinuity of the semigroups seem to fail here; also, there is little hope of obtaining a Frechet space, even when I is countable (see Theorem 27.5).

Remark 27.4. Informally, $\mathcal{D}(\mathcal{A})$ may be described in terms of the abstract Cauchy problem, as follows. In order that x_0 be in $\mathcal{D}(\mathcal{A})$, one wants to be able to first find a mild solution, u_1, of (0.1), for $A = A_{\alpha_1}$, $x = x_0$; then, for $x = x_1 \equiv u_1(t_1, x_0)$, one wants a mild solution u_2, of (0.1), for $A = A_{\alpha_2}$, etc. This means we are following one solution curve, then starting another solution curve at an arbitrary point on the original solution curve, and continuing this process indefinitely.

Theorem 27.5. *Suppose that, for all $\alpha \in I$, the abstract Cauchy problem (0.1), with $A = A_\alpha, x = 0$, has no nontrivial solutions. Then*

(1) $Z(\mathcal{A})$ *is a sequentially complete locally convex space;*

(2) $Z(\mathcal{A}) \hookrightarrow X$;

(3) *for all $\alpha \in I, A_\alpha|_{Z(\mathcal{A})}$ generates a strongly continuous locally equicontinuous semigroup; and*

(4) $Z(\mathcal{A})$ *is maximal-unique, that is, if Y is a locally convex space satisfying (2) and (3), then $Y \hookrightarrow Z(\mathcal{A})$.*

179

If I is countable, then $Z(\mathcal{A})$ is a Frechet space.

Proof: Properties (2) and (3) are clear from the definition of $Z(\mathcal{A})$ (Definition 27.2); for equicontinuity, note that, for $a_0 \le t_0 \le b_0$,

$$\|e^{t_0 A_{\alpha_0}} x\|_{\mathcal{A}, \vec{a}, \vec{b}, \vec{\alpha}}$$

$$= \sup\{\|(\prod_{k=0}^{n} e^{t_k A_{\alpha_k}}) x\| \mid t_k \in [a_k, b_k] \, (1 \le k \le n)\} \le \|x\|_{\mathcal{A}, \vec{a_1}, \vec{b_1}, \vec{\alpha_1}},$$

where $\vec{a_1} \equiv\, < a_0, a_1, \dots a_n >, \vec{b_1} \equiv\, < b_0, b_1, \dots b_n >, \vec{\alpha_1} \equiv\, < \alpha_k >_{k=0}^{n}$.

To prove (1), suppose $< x_n >$ is a Cauchy sequence in $Z(\mathcal{A})$. Then there exists $x \in X$ such that $x_n \to x$ in X. We will show, by induction on m, the number of coordinates in $\vec{t}$ and $\vec{\alpha}$, that for all finite sequences $\vec{t}$ and $\vec{\alpha}$,

$$x \in \mathcal{D}\left(\prod_{k=1}^{m} e^{t_k A_{\alpha_k}}\right),$$

and

$$\lim_{j \to \infty} \|(\prod_{k=1}^{m} e^{t_k A_{\alpha_k}})(x_j - x)\| = 0. \qquad (27.6)$$

For $m = 0$ (27.6) clearly holds. Suppose (27.6) is valid for all sequences of length m. Given $t_1, \dots, t_{m+1} \ge 0, \alpha_1, \dots, \alpha_{m+1} \in I, j \in \mathbf{N}$, let

$$B \equiv A_{\alpha_{m+1}}, \; y_j \equiv (\prod_{k=1}^{m} e^{t_k A_{\alpha_k}}) x_j, \; y \equiv (\prod_{k=1}^{m} e^{t_k A_{\alpha_k}}) x.$$

Since $< x_n >$ is Cauchy in $Z(\mathcal{A})$, the functions $t \mapsto e^{tB} y_j$, from $[0, \infty)$ into X, are uniformly Cauchy on compact subsets of $[0, \infty)$, thus converge uniformly on compact subsets to a continuous $\phi : [0, \infty) \to X$. Note that $\phi(0) = y$. For any $j \in \mathbf{N}, t \ge 0$,

$$B\left(\int_0^t e^{sB} y_j \, ds\right) = e^{tB} y_j - y_j;$$

thus, since B is closed and the convergence of $e^{sB} y_j$ is uniform on $[0, t]$,

$$B\left(\int_0^t \phi(s) \, ds\right) = \phi(t) - y.$$

Because of the uniqueness of the mild solutions of (0.1), this is saying that $y \in \mathcal{D}(e^{tB})$, and $e^{tB} y = \phi(t)$, so that $e^{tB}(y_j - y) \to 0$, for all $t \ge 0$. This concludes the induction, proving (27.6).

To see that $\|x\|_{A,\vec{a},\vec{b},\vec{\alpha}}$ is finite, for any $\vec{a},\vec{b},\vec{\alpha}$, first note that, since $<x_n>$ is Cauchy, there exists a constant $K_{\vec{a},\vec{b},\vec{\alpha}}$ such that $\|x_n\|_{A,\vec{a},\vec{b},\vec{\alpha}} \leq K_{\vec{a},\vec{b},\vec{\alpha}}$ for all $n \in \mathbf{N}$. Thus, by (27.6), for any $t_k \in [a_k,b_k]$,

$$\|(\prod_{k=1}^{m} e^{t_k A_{\alpha_k}})x\| = \lim_{j\to\infty} \|(\prod_{k=1}^{m} e^{t_k A_{\alpha_k}})x_j\| \leq K_{\vec{a},\vec{b},\vec{\alpha}},$$

so that $\|x\|_{A,\vec{a},\vec{b},\vec{\alpha}} \leq K_{\vec{a},\vec{b},\vec{\alpha}} < \infty$.

Next, we will show that $\|(x_n - x)\|_{A,\vec{a},\vec{b},\vec{\alpha}}$ converges to 0, as $n \to \infty$. Fix $\epsilon > 0$ and choose N so that $\|(x_n - x_j)\|_{A,\vec{a},\vec{b},\vec{\alpha}} < \epsilon$, for all $n,j > N$. For $n > N$,

$$\|(\prod_{k=1}^{m} e^{t_k A_{\alpha_k}})(x_n - x)\| = \lim_{j\to\infty} \|(\prod_{k=1}^{m} e^{t_k A_{\alpha_k}})(x_n - x_j)\|,$$

for any $t_k \in [a_k,b_k]$, by (27.6), thus is less than or equal to ϵ. Taking suprema, we conclude that $\|(x_n - x)\|_{A,\vec{a},\vec{b},\vec{\alpha}} \leq \epsilon$, for all $n > N$.

To prove (1), all that remains is to show that x satisfies (2) of Definition 27.2. Fix $\vec{a}, \vec{b}, \vec{\alpha}$. From the definitions of the seminorms, it is not hard to see that, for $\alpha \in I$, the maps $t \mapsto e^{tA_\alpha}x_n$, from $[0,\infty)$ into $(\mathcal{D}(A),\| \ \|_{A,\vec{a},\vec{b},\vec{\alpha}})$, converge uniformly to $e^{tA_\alpha}x$, as $n \to \infty$, on any bounded subset of $[0,\infty)$. Since $x_n \in Z(A)$, for all n, this implies that the map $t \mapsto e^{tA_\alpha}x$, from $[0,\infty)$ into $(\mathcal{D}(A),\| \ \|_{A,\vec{a},\vec{b},\vec{\alpha}})$, is continuous, as desired.

For (4), suppose Y is a locally convex space, topologized by the seminorms $\{\| \ \|_\beta\}_{\beta \in J}$, satisfying (2) and (3).

Since $Y \hookrightarrow X$, there exist $\beta_1,...,\beta_n \in J, c_1,...c_n \geq 0$ such that

$$\|x\| \leq \sum_{i=1}^{n} c_i \|x\|_{\beta_i}, \forall x \in Y. \tag{27.7}$$

By the local equicontinuity, for any $\vec{a},\vec{b},\vec{\alpha}$, for $1 \leq i \leq n$, there exists $\{\beta_{i,j}\}_{j=1}^{N(i)} \subseteq J, \{d_{i,j}\}_{j=1}^{N(i)} \subseteq [0,\infty)$ such that

$$\sup\{\|(\prod_{k=1}^{n} e^{t_k A_{\alpha_k}}|_Y)x\|_{\beta_i} \mid t_k \in [a_k,b_k]\} \leq \sum_{j=1}^{N(i)} d_{i,j}\|x\|_{\beta_{i,j}}, \forall x \in Y. \tag{27.8}$$

Since $Y \hookrightarrow X$, it follows that $Y \subseteq \mathcal{D}(A)$, with

$$\left(\prod_{k=1}^{n} e^{t_k A_{\alpha_k}}|_Y\right)x = \left(\prod_{k=1}^{n} e^{t_k A_{\alpha_k}}\right)x, \tag{27.9}$$

181

for all $x \in Y$.

Assertions (27.7), (27.8) and (27.9) now imply that

$$\|x\|_{\mathcal{A},\vec{a},\vec{b},\vec{\alpha}} \leq \sum_{i,j} c_i d_{i,j} \|x\|_{\beta_{i,j}}, \forall x \in Y;$$

this is saying that $Y \hookrightarrow Z(\mathcal{A})$, once we establish that (2) of Definition 27.2 is satisfied for any $x \in Y$.

Fix $\vec{a}, \vec{b}, \vec{\alpha}, \alpha_0 \in I$. Since $\{e^{tA_\alpha|_Y}\}_{t \geq 0}$ is a strongly continuous semi-group, for all $\alpha \in I$, it follows that, for any $\beta \in J$, the map

$$t_0 \mapsto \sup\{\| \left(\prod_{k=0}^{n} e^{t_k A_{\alpha_k}} \right) x\|_\beta \,|\, t_k \in [a_k, b_k] (1 \leq k \leq n)\}$$

is continuous. Since $Y \hookrightarrow X$, the map $t \mapsto e^{tA_{\alpha_0}}x$, from $[0, \infty)$ into $(\mathcal{D}(\mathcal{A}), \| \ \|_{\mathcal{A},\vec{a},\vec{b},\vec{\alpha}})$ is continuous, as desired.

Finally, if I is countable, then we may, in the definitions of the semi-norms in Definition 27.2, restrict ourselves to $\vec{a}, \vec{b}$ equal to finite sequences of rational numbers, so that $Z(\mathcal{A})$ is a sequentially complete locally convex space topologized by a countable family of seminorms, making it a Frechet space. ∎

XXVIII. EXPONENTIALLY BOUNDED SIMULTANEOUS SOLUTION SPACE

Let $\mathcal{A} \equiv \{A_\alpha\}_{\alpha \in I}$ be, as in the previous chapter, an arbitrary family of closed operators. In this chapter, we wish to construct a maximal continuously embedded *Banach* space, Z, such that $A_\alpha|_Z$ generates a strongly continuous semigroup, for all α.

We need, as in Chapter V, to restrict ourselves to exponentially bounded solutions. For a closed operator A, we will denote by $Z_{exp}(A)$ the *exponentially bounded solution space* for A, the set of all x in $Z(A)$ (see Terminology 27.1) for which the mild solution of the abstract Cauchy problem (0.1) $t \mapsto u(t,x)$, from $[0,\infty)$ into X, is exponentially bounded.

Definition 28.1. Suppose A is a closed operator such that $(r - A)$ is injective, for r large; that is, there exists $r_0 \in \mathbf{R}$ such that $(r - A)$ is injective, for all $r > r_0$. By Proposition 2.9, for any $x \in Z_{exp}(A)$, the exponentially bounded mild solution of (0.1) is unique.

We then define, for $t \geq 0$, the operator $(e^{tA})_{exp}$ by

$$(e^{tA})_{exp}x \equiv u(t,x), \quad \mathcal{D}((e^{tA})_{exp}) \equiv Z_{exp}(A).$$

Definition 28.2. We will write $Z_{exp}(\mathcal{A})$ for the *exponentially bounded simultaneous solution space* for $\mathcal{A}$

$$Z(\mathcal{A}) \bigcap \mathcal{D}(\prod_{k=1}^{n}(e^{t_k A_{\alpha_k}})_{exp}),$$

over all finite sequences $<t_k>_{k=1}^n, <\alpha_k>_{k=1}^n$.

Definition 28.3. We will write $|\vec{t}| \equiv \sum_{k=1}^n t_k$.

For $x \in Z_{exp}(\mathcal{A}), \omega \in \mathbf{R}$, define

$$\|x\|_{\mathcal{A},\omega} \equiv \sup\{\|e^{-\omega|\vec{t}|}(\prod_{k=1}^{n} e^{t_k A_{\alpha_k}})x\| \,|\, t_k \geq 0, \alpha_k \in I, (n-1) \in \mathbf{N}\}.$$

The space $Z_\omega(\mathcal{A})$ is defined to be the set of all $x \in Z_{exp}(\mathcal{A})$ such that

(1) $\|x\|_{\mathcal{A},\omega} < \infty$; and

(2) The map $s \mapsto e^{sA_\alpha}x$, from $[0,\infty)$ into $(Z_{exp}(\mathcal{A}), \| \; \|_{\mathcal{A},\omega})$ is continuous, for all $\alpha \in I$.

We will also write $Z_\omega(\mathcal{A})$ for the normed vector space $(Z_\omega(\mathcal{A}), \| \ \|_{\mathcal{A},\omega})$

Remark 28.4. When $\mathcal{A} = \{A\}$, a single operator, then $Z_0(\mathcal{A})$ equals the set of all x for which (0.1) has a bounded, uniformly continuous mild solution (see Chapter V).

Theorem 28.5. *Suppose that, for all $\alpha \in I$, there exists $r_\alpha \in \mathbf{R}$ such that $(r - A_\alpha)$ is injective, for all $r > r_\alpha$. Then, for any real ω,*

(1) *$Z_\omega(\mathcal{A})$ is a Banach space;*

(2) *$Z_\omega(\mathcal{A}) \hookrightarrow X$;*

(3) *for all $\alpha \in I, A_\alpha|_{Z_\omega(\mathcal{A})}$ generates a strongly continuous semigroup, with $\|e^{tA_\alpha}|_{Z_\omega(\mathcal{A})}\| \leq e^{\omega t}$, for all $t \geq 0$; and*

(4) *Z is maximal-unique, that is, if Y is a normed vector space satisfying (2) and (3), then $Y \hookrightarrow Z$.*

Example 28.6. It is clear, by maximality, that $Z_0(\{A_1, A_2\})$ is contained in $Z_0(A_1) \bigcap Z_0(A_2)$, for any pair of closed operators A_1, A_2.

We give here an example of commuting closed operators A_1 and A_2 such that $Z_0(\{A_1, A_2\}) \neq Z_0(A_1) \bigcap Z_0(A_2)$. This will be a two dimensional version of the "bumpy translation" space of Example 4.17.

For Ω a region in $\mathbf{R}^2$, write $BUC(\Omega)$ for the set of all bounded uniformly continuous complex-valued functions on Ω, $BUC^1(\Omega)$ for the set of all $f \in BUC(\Omega)$ such that $\frac{\partial f}{\partial x}$ and $\frac{\partial f}{\partial y}$ exist and are in $BUC(\Omega)$.

Let $A_1 \equiv \frac{\partial}{\partial x}$, the generator of left-translation,

$$(e^{tA_1}f)(x,y) \equiv f(x+t, y),$$

$A_2 \equiv \frac{\partial}{\partial y}$, the generator of downward translation,

$$(e^{tA_2}f)(x,y) \equiv f(x, y+t),$$

both with maximal domains.

We let both these operators act on a space where translation is not strongly continuous, $X \equiv BUC(\mathbf{R}^2) \bigcap BUC^1(\{(x,y) \in \mathbf{R}^2 \,|\, x,y \leq 0\})$.

Then it is not hard to see that $Z_0(\{A_1, A_2\}) = BUC^1(\mathbf{R}^2)$, while $Z_0(A_1) \bigcap Z_0(A_2) = BUC(\mathbf{R}^2) \bigcap BUC^1(\{(x,y) \in \mathbf{R}^2 \,|\, x \leq 0 \text{ or } y \leq 0\})$.

Suppose that $(\omega, \infty) \subseteq \rho(A_\alpha)$, for all $\alpha \in I$. We may then use the resolvents to construct $Z_\omega(\mathcal{A})$. Without loss of generality, suppose $\omega = 0$. Then we may mimic the construction and proof in [K2] as follows.

184

Let $Y(\mathcal{A})$ be the set of all $x \in X$ such that

$$\|x\|_{Y(\mathcal{A})} \equiv \sup\{\|(\prod_{k=1}^{n} \lambda_k(\lambda_k - A_{\alpha_k})^{-1})x\| \mid \lambda_k > 0, \alpha_k \in I, (n-1) \in \mathbf{N}\}$$

is finite.

Then, using the maximality of our construction in this chapter, we have the following.

Theorem 28.7. *Suppose* $(0,\infty) \subseteq \rho(A_\alpha)$, *for all* $\alpha \in I$.

(1) $Z_0(\mathcal{A})$ *equals the closure, in* $Y(\mathcal{A})$, *of* $\bigcap \mathcal{D}(A_{\alpha_k}|_{Y(\mathcal{A})})$, *where the intersection is taken over all finite sequences* $< \alpha_k > \subseteq I$.

(2) $\|x\|_{Z_0(\mathcal{A})} = \|x\|_{Y(\mathcal{A})}$, *for all* $x \in Z_0(\mathcal{A})$.

We should mention that, for many interesting examples, the resolvent condition of this section will not be satisfied; for example, the matrices of operators in Chapter XXX.

Proof of Theorem 28.5: Without loss of generality, we may assume $\omega = 0$ (otherwise work with $(A_\alpha - \omega)$).

Properties (2) and (3) are clear from the definition of $\| \ \|_{\mathcal{A},0}$ and (2) of Definition 28.3.

To prove (1), suppose $< x_n >$ is a Cauchy sequence in $Z_0(\mathcal{A})$. The same argument as in the proof of Theorem 27.5, with uniform convergence on compact subsets of $[0,\infty)$ replaced by uniform convergence on $[0,\infty)$, shows that there exists

$$x \in \bigcap \mathcal{D}\left(\prod_{k=1}^{m} e^{t_k A_{\alpha_k}}\right),$$

with the intersection taken over all finite sequences $\vec{t}, \vec{\alpha}$, such that $\|(x_n - x)\|_{\mathcal{A},0}$ converges to 0, as $n \to \infty$.

All that remains is to show that x satisfies (2) of Definition 28.3. It is clear from the definition of the norm that $\|e^{sA_\alpha}(x_n - x)\|_{\mathcal{A},0} \leq \|(x_n - x)\|_{\mathcal{A},0}$, for all $s \geq 0, n \in \mathbf{N}, \alpha \in I$. Thus the functions $s \mapsto e^{sA_\alpha}x_n$, from $[0,\infty) \to (Z_{exp}(\mathcal{A}), \| \ \|_{\mathcal{A},0})$, converge uniformly to $e^{sA_\alpha}x$. Since $x_n \in Z_0(\mathcal{A})$, for all n, this implies that $s \mapsto e^{sA_\alpha}x$, from $[0,\infty)$ into $(Z_{exp}(\mathcal{A}), \| \ \|_{\mathcal{A},0})$, is continuous, as desired.

For (4), suppose Y is a normed vector space satisfying (2) and (3). There exists a constant M such that $\|x\| \leq M\|x\|_Y$, for all $x \in Y$. For any $\vec{t}, \vec{\alpha}, x \in Y$,

$$\|(\prod_{k=1}^{m} e^{t_k A_{\alpha_k}})x\| \leq M\|(\prod_{k=1}^{m} e^{t_k A_{\alpha_k}}|_Y)x\|_Y \leq M\|x\|_Y.$$

185

Thus, $\|x\|_{\mathcal{A},0} \leq M\|x\|_Y$. To show that $Y \hookrightarrow Z_0(\mathcal{A})$, all that remains is to show, for any $x \in Y$, that x satisfies (2) of Definition 28.3. The map $s \mapsto e^{sA_\alpha}x$, from $[0,\infty) \to Y$, is continuous, thus since $\|x\|_{\mathcal{A},0} \leq M\|x\|_Y$, the same is true as a map from $[0,\infty)$ into $(Z_0(\mathcal{A}), \| \ \|_{\mathcal{A},0})$. ∎

XXIX. SIMULTANEOUS EXISTENCE FAMILIES

Let $\mathcal{A} \equiv \{A\}_\alpha$ be as in Terminology 27.1.

As with a single operator, we need simple criteria for determining what $Z_\omega(\mathcal{A})$ and $Z(\mathcal{A})$ contain. In this chapter, we introduce a multivariable version of C-existence families (see Chapter II), *simultaneous C-existence families*, that is equivalent to $Z(\mathcal{A})$ containing the image of a bounded operator C. Exponentially bounded simultaneous C-existence families will provide the same equivalence for $Z_\omega(\mathcal{A})$.

The spaces $Z(\mathcal{A})$ and $Z_\omega(\mathcal{A})$ will be difficult to calculate exactly, in general. We need a method for approximating the sets and their topologies. This will be provided by our simultaneous C-existence families.

Corresponding to a Banach subspace, we give necessary and sufficient Hille-Yosida type theorems for the existence of an exponentially bounded simultaneous existence family.

When C commutes with A_α, for all $\alpha \in I$, this is a multivariable version of a C-regularized semigroup (see Chapter III).

Definition 29.1. If $Im(C) \subseteq Z(\mathcal{A})$ and

$$W(\vec{t}, \vec{\alpha}) \equiv \left(\prod_k e^{t_k A_{\alpha_k}} \right) C$$

is a bounded operator from X to itself, for all finite sequences $\vec{t}, \vec{\alpha}$, then $\{W(\vec{t}, \vec{\alpha})\}_{t_k \geq 0, \alpha_k \in I}$ is a *simultaneous C-existence family* for $\mathcal{A}$.

Remark 29.2. By Banach-Steinhaus and the definition of $Z(\mathcal{A})$ (see Definition 27.2(1)), W is uniformly bounded for t_k contained in bounded subsets of $[0, \infty)$; that is, for any finite sequences $\vec{a}, \vec{b}, \vec{\alpha}$, there exists a constant $M_{\vec{a}, \vec{b}, \vec{\alpha}}$ such that

$$\|W(\vec{t}, \vec{\alpha})\| \leq M_{\vec{a}, \vec{b}, \vec{\alpha}}, \ \forall t_k \in [a_k, b_k];$$

in fact, Theorem 29.4 will show that the boundedness of $W(\vec{t}, \vec{\alpha})$ follows automatically from the fact that $Im(C) \subseteq Z(\mathcal{A})$.

We need to define exponential boundedness.

Definition 29.3. We will say that the simultaneous C-existence family from Definition 29.1 is $O(e^{\omega |\vec{t}|})$ if there exists a constant M such that

$$\|W(\vec{t}, \vec{\alpha})\| \leq M e^{\omega |\vec{t}|},$$

for all $\vec{t}, \vec{\alpha}$.

In the following theorems, we will write $[Im(C)]$ for the Banach space $Im(C)$ with norm $\|y\|_{[Im(C)]} \equiv \inf\{\|x\| \mid Cx = y\}$.

Theorem 29.4. *Suppose that, for all $\alpha \in I$, there are no nontrivial local solutions of the abstract Cauchy problem* (0.1) *with $x = 0, A \equiv A_\alpha$. Then the following are equivalent.*

(a) $Im(C) \subseteq Z(\mathcal{A})$.

(b) $[Im(C)] \hookrightarrow Z(\mathcal{A})$.

(c) *There exists a simultaneous C-existence family for $\mathcal{A}$.*

Theorem 29.5. *Suppose that $\omega \in \mathbb{R}$ and $(\lambda - A_\alpha)$ is injective, for all $\alpha \in I, \lambda > \omega$. Then the following are equivalent.*

(a) $Im(C) \subseteq Z_\omega(\mathcal{A})$.

(b) $[Im(C)] \hookrightarrow Z_\omega(\mathcal{A})$.

(c) *There exists a simultaneous C-existence family for $\mathcal{A}$ that is $O(e^{\omega|\vec{t}|})$.*

If $\bigcap_{\alpha \in I} \mathcal{D}(A_\alpha)$ is dense in X, these are equivalent to

(d) $Im(C) \subseteq Im\left(\prod_{k=1}^n (\lambda_k - A_{\alpha_k})\right)$, *for all finite sequences $\{\lambda_k\}_{k=1}^n \subseteq (\omega, \infty), \{\alpha_k\}_{k=1}^n \subseteq I$ and there exists a constant M such that*

$$\left\| \left(\prod_{k=1}^n (\lambda_k - \omega)(\lambda_k - A_{\alpha_k})^{-1} \right) C \right\| \leq M,$$

for any $\{\lambda_k\}_{k=1}^n \subseteq (\omega, \infty), \{\alpha_k\}_{k=1}^n \subseteq I, n - 1 \in \mathbb{N}$.

Proof of Theorem 29.4: (c) $\to$ (b). For any finite sequences $\vec{a}, \vec{b}, \vec{\alpha}$, there exists a constant $M_{\vec{a}, \vec{b}, \vec{\alpha}}$ such that

$$\|W(\vec{t}, \vec{\alpha})\| \leq M_{\vec{a}, \vec{b}, \vec{\alpha}}, \; \forall t_k \in [a_k, b_k].$$

(See Remark 29.2.) Then it's clear that, for any $\vec{a}, \vec{b}, \vec{\alpha}$, $y = Cx, x \in X$,

$$\|y\|_{\mathcal{A}, \vec{a}, \vec{b}, \vec{\alpha}} = \|Cx\|_{\mathcal{A}, \vec{a}, \vec{b}, \vec{\alpha}} = \sup\{\|W(\vec{t}, \vec{\alpha})x\| \,|\, t_k \in [a_k, b_k]\} \leq M_{\vec{a}, \vec{b}, \vec{\alpha}}\|x\|,$$

so that, by taking infima on the right, we have

$$\|y\|_{\mathcal{A}, \vec{a}, \vec{b}, \vec{\alpha}} \leq M_{\vec{a}, \vec{b}, \vec{\alpha}}\|y\|_{[Im(C)]},$$

for all $y \in Im(C)$, as desired.

(a) $\to$ (c). We must show that

$$W(\vec{t}, \vec{\alpha})(x) \equiv \left(\prod_{k=1}^m e^{t_k A_{\alpha_k}} \right) Cx$$

188

defines a bounded operator on X, for all $\vec{t}, \vec{\alpha}$.

We will show this by induction on m, the number of coordinates in $\vec{t}$ and $\vec{\alpha}$, where $\prod_{k=1}^{0} \equiv I$. Suppose $W(\vec{t}, \vec{\alpha})$ is bounded, for vectors with $m-1$ coordinates ($m \in \mathbf{N}$). Given $\vec{t}, \vec{\alpha}$ with m coordinates, $T \geq 0$, define $W : X \to C([0, T], X)$ by

$$(Wx)(t_m) \equiv \left(\prod_{k=1}^{m} e^{t_k A_{\alpha_k}} \right) Cx \quad (0 \leq t_m \leq T).$$

Let $G \equiv \left(\prod_{k=1}^{m-1} e^{t_k A_{\alpha_k}} \right) C, B \equiv A_{\alpha_m}$.

Suppose $x_n \to x$ and $Wx_n \to \phi$, as $n \to \infty$. Then, by the induction hypothesis, $Gx_n \to Gx$. For any $0 \leq t \leq T$, since B is closed, and for all n,

$$B \left(\int_0^t Wx_n(s) \, ds \right) = Wx_n(t) - Gx_n,$$

it follows that $\int_0^t \phi(s) \, ds \in \mathcal{D}(B)$, with

$$B \left(\int_0^t \phi(s) \, ds \right) = \phi(t) - Gx. \tag{29.6}$$

Also note that $\phi(0) = \lim_{n \to \infty} Wx_n(0) = \lim_{n \to \infty} Gx_n = Gx$. By (29.6), this implies that $\phi(t) = e^{tB} Gx = Wx(t)$.

Thus W is closed, hence, by the closed graph theorem, bounded, which clearly implies that $W(\vec{t}, \vec{\alpha})$ is bounded, for vectors $\vec{t}, \vec{\alpha}$ with m coordinates. This completes the induction. ∎

Proof of Theorem 29.5: By translating, we may assume that $\omega = 0$.
(c) $\to$ (b). There exists a constant M such that

$$\left\| \left(\prod_{k=1}^{n} e^{t_k A_{\alpha_k}} \right) C \right\| \leq M,$$

for any finite sequences $\vec{t}, \vec{\alpha}$. Thus, for $y \in Im(C), Cx = y$,

$$\left\| \left(\prod_{k=1}^{n} e^{t_k A_{\alpha_k}} \right) y \right\| \leq M \|x\|,$$

so that taking suprema on the left and infima on the right implies that $\|y\|_{A,0} \leq M \|y\|_{[Im(C)]}$, as desired.

189

(a) → (c). As in the proof of Theorem 29.4, with $[0,T]$ replaced by $[0,\infty)$ and $C([0,T],X)$ replaced by $BC([0,\infty),X)$,

$$W(\vec{t},\vec{\alpha})(x) \equiv \left(\prod_{k=1}^{m} e^{t_k A_{\alpha_k}} \right) Cx$$

defines a bounded operator on X, for all finite sequences $\vec{t},\vec{\alpha}$. By the Banach-Steinhaus theorem, since $Im(C) \subseteq Z_0(\mathcal{A})$, $\|W(\vec{t},\vec{\alpha})\|$ is uniformly bounded.

Now suppose that $\bigcap_{\alpha \in I} \mathcal{D}(A_\alpha)$ is dense. We will write Z for $Z_0(\mathcal{A})$.

(b) → (d). This follows from the Hille-Yosida theorem, applied to $A_\alpha|_Z$. First, for $\lambda_k > 0, \alpha_k \in I$, since $(\lambda - A_{\alpha_k}|_Z)$ is surjective, it follows that $Im(C) \subseteq Im(\prod_{k=1}^{n}(\lambda_k - A_{\alpha_k}))$, for all finite sequences $\{\lambda_k\}_{k=1}^{n} \subseteq (0,\infty), \{\alpha_k\}_{k=1}^{n} \subseteq I$.

There exists a constant M such that $\|y\|_Z \leq M\|y\|_{[Im(C)]}$, for all $y \in Im(C)$. Thus for any $x \in X, \lambda_k > 0, \alpha_k \in I$,

$$\left\| \left(\prod_{k=1}^{n} \lambda_k(\lambda_k - A_{\alpha_k})^{-1} \right) Cx \right\| \leq \left\| \left(\prod_{k=1}^{n} \lambda_k(\lambda_k - A_{\alpha_k})^{-1} \right) Cx \right\|_Z$$

$$\leq \|Cx\|_Z \leq M\|Cx\|_{[Im(C)]} \leq M\|x\|.$$

(d) → (a). Define, for $x \in \bigcap\{Im(\prod_{k=1}^{n}(\lambda_k - A_{\alpha_k})) \,|\, \lambda_k > 0, \alpha_k \in I, n \in \mathbf{N}\}$,

$$\|x\|_W \equiv \sup\{\left\| \left(\prod_{k=1}^{n} \lambda_k(\lambda_k - A_{\alpha_k})^{-1} \right) x \right\| \,|\, \lambda_k > 0, \alpha_k \in I, n-1 \in \mathbf{N}\},$$

and let W be the normed vector space of all x such that $\|x\|_W < \infty$. Let $\tilde{W}$ be the completion, with respect to $\|\ \|_W$, of W. Note that $\tilde{W} \hookrightarrow X$. Let Y be the closure, in $\tilde{W}$, of $\bigcap_{\alpha \in I} \mathcal{D}(A_\alpha|_{\tilde{W}})$. Then for all $\alpha \in I$, by the Hille-Yosida theorem, $A_\alpha|_Y$ generates a strongly continuous semigroup of contractions.

By Theorem 28.5(4), $Y \hookrightarrow Z$. By (d), $[Im(C)] \hookrightarrow W$, hence $[Im(C)] \hookrightarrow \tilde{W}$. Let $\mathcal{D} \equiv \bigcap_{\alpha \in I} \mathcal{D}(A_\alpha)$. Then $[C(\mathcal{D})] \hookrightarrow Y$. For $x \in X$, choose $x_n \in \mathcal{D}$ such that $x_n \to x$. Then $Cx_n \to Cx$, in $[Im(C)]$, hence in $\tilde{W}$. Thus $[Im(C)] \hookrightarrow Y$. ∎

XXX. SIMULTANEOUS EXISTENCE FAMILIES FOR MATRICES OF OPERATORS

In this chapter, we will show that many families of matrices of constant coefficient differential operators have a simultaneous C-existence family, for simple and explicit choices of C, with dense image. More generally, as in Chapters XII–XIV, we will consider matrices of polynomials of commuting generators of bounded strongly continuous groups.

Our construction yields maximal continuously embedded Banach subspaces where large classes of systems of partial differential equations, including symmetric hyperbolic systems on $L^p(\mathbf{R}^n)(1 \leq p < \infty)$ or $C_0(\mathbf{R}^n)$, are simultaneously well-posed, that is, the corresponding matrices of operators all generate strongly continuous semigroups (see Example 30.7).

To obtain exponentially bounded simultaneous existence families (hence a Banach subspace on which the operators generate strongly continuous semigroups), we need the numerical range of the corresponding complex-valued matrices to have real part bounded above (see Theorems 30.2–30.6). We also show that the collection of *all* matrices of polynomials of commuting generators of bounded strongly continuous groups has a simultaneous existence family (Theorem 30.9).

We use the same terminology as in Chapters XII–XIV.

Definition 30.1. $\mathcal{F}(\omega)$ will be the set of all $m \times m$ matrices of polynomials in n variables $(p_{i,j})_{i,j=1}^m$ such that

$$\sup\{Re(z) \mid z \in n.r.(p_{i,j}(x))_{i,j=1}^m\} \leq \omega,$$

for all $x \in \mathbf{R}^n$.

$\mathcal{F}(\omega, N)$ will be the subset of $\mathcal{F}(\omega)$ such that the degree of $p_{i,j}$ is less than or equal to N, for all i,j.

Finally, we define $\mathcal{F}(\omega, N, M)$ to be the set of all $f \in \mathcal{F}(\omega, N)$ such that

$$|D^\alpha p_{i,j}(x)| \leq M(1+|x|^2)^{\frac{1}{2}(N-|\alpha|)},$$

for all $x \in \mathbf{R}^n, 1 \leq i,j \leq m, |\alpha| \leq k$.

Note that, for $m = 1$, $\mathcal{F}(\omega)$ is the set of all polynomials in n variables, p, such that $Re(p(x)) \leq \omega$, for all $x \in \mathbf{R}^n$.

To get an exponentially bounded simultaneous existence family, we need uniform growth conditions on the polynomials (Theorems 30.2 and 30.4). When these are removed, we still obtain continuously embedded Banach subspaces on which all operators generate strongly continuous semigroups (Theorems 30.3, 30.5 and 30.6).

Theorem 30.2. For any $M > 0$, let

$$\mathcal{A}_M \equiv \{(p_{i,j}(B)) \mid (p_{i,j}) \in \mathcal{F}(\omega, N, M)\}.$$

Suppose $r > \frac{k}{2}(N-1) + \frac{n}{4}$.

Then for all $\mu > \omega, M > 0$, there exists a simultaneous $(1 + |B|^2)^{-r}$-existence family for $\mathcal{A}_M$ that is $O(e^{\mu|\vec{t}|})$.

Theorem 30.3. Let

$$\mathcal{A} \equiv \{(p_{i,j}(B)) \mid (p_{i,j}) \in \mathcal{F}(\omega, N)\}.$$

Suppose $r > \frac{k}{2}(N-1) + \frac{n}{4}$.

Then there exists a Banach space W such that

$$([\mathcal{D}(|B|^r)])^m \hookrightarrow W \hookrightarrow X^m,$$

and for all $A \in \mathcal{A}$, $A|_W$ generates a strongly continuous semigroup.

Theorem 30.4. For any $M > 0$, let $\mathcal{A}_M$ be the set of all $(p_{i,j}(B))$ such that $(p_{i,j}) \in \mathcal{F}(\omega)$ and

$$|D^\alpha p_{i,j}(x)| \le M e^{|x|}, \forall x \in \mathbf{R}^n, 1 \le i, j \le m, |\alpha| \le k.$$

Then for all $\mu > \omega, M > 0$, there exists a simultaneous $e^{-|B|}$-existence family for $\mathcal{A}_M$ that is $O(e^{\mu|\vec{t}|})$.

Theorem 30.5. Let $\mathcal{A} \equiv \{(p_{i,j}(B)) \mid (p_{i,j}) \in \mathcal{F}(\omega)\}$.

Then there exists a Banach space W such that

$$([Im(e^{-|B|})])^m \hookrightarrow W \hookrightarrow X^m,$$

and for all $A \in \mathcal{A}$, $A|_W$ generates a strongly continuous semigroup.

Theorem 30.6. Let

$$\mathcal{A} \equiv \{(p_{i,j}(B))_{i,j=1}^m \mid (p_{i,j}(x))_{i,j=1}^m \text{ is symmetric for all } x \in \mathbf{R}^n\}.$$

Then there exists a Banach space W such that

$$([Im(e^{-|B|})])^m \hookrightarrow W \hookrightarrow X^m,$$

and for all $A \in \mathcal{A}, \ell \in \mathbf{N}, i(A|_W)^\ell$ generates a strongly continuous group.

Example 30.7. Note that $\mathcal{A}$, from Theorem 30.6, includes all symmetric hyperbolic systems, that is, operators A of the form

$$A = \sum_{j=1}^{n} \mathcal{M}_j D_j,$$

for some constant $m \times m$ complex-valued symmetric matrices $\mathcal{M}_j$, on $(L^p(\mathbf{R}^n))^m (1 \leq p < \infty), C_0(\mathbf{R}^n)^m$, etc. More generally, it includes operators of the form

$$A = \sum_{j=1}^{n} \mathcal{M}_j D_j + \mathcal{M}_0, \tag{30.8}$$

for some constant $m \times m$ complex-valued symmetric matrices $\mathcal{M}_j (0 \leq j \leq n)$, by choosing

$$(p_{i,j}(x))_{i,j=1}^{m} \equiv \sum_{j=1}^{n} \mathcal{M}_j x_j + \mathcal{M}_0,$$

for $x \in \mathbf{R}^n$.

Thus, on the dense Banach subspace W, of $(L^p(\mathbf{R}^n))^m (1 \leq p < \infty)$, or $(C_0(\mathbf{R}^n))^m$, of Theorem 30.6, all the abstract Cauchy problems (1.1) corresponding to operators as in (30.8), are simultaneously well-posed. This includes the wave equation, Maxwell's equations, the abstract Cauchy problem corresponding to the Dirac operator, and many other famous partial differential equations; see [Go2, Section II.9] and [Gi-S] for more examples.

If we remove the desire for exponential boundedness (hence replace a Banach subspace by a Frechet subspace), we obtain a much more sweeping statement.

Theorem 30.9. *The collection of all matrices of polynomials of B has a simultaneous $e^{-|B|}$-existence family.*

In particular, all higher order problems of the form

$$u^{(n)}(t) = \sum_{\ell=0}^{n-1} p_\ell(B) u^{(\ell)}(t) \ (t \geq 0), \ u^{(\ell)}(0) = x_\ell \ (0 \leq \ell \leq n-1);$$

this includes constant-coefficient partial differential equations of arbitrarily high order; are well-posed in the sense of Frechet spaces.

We will need the following generalization of Lemma 13.13.

Lemma 30.10. *There exists finite K such that*

$$\left\| \left((\prod_{\ell} e^{(p_{i,j,\ell}(x))_{i,j=1}^m}) g(x) \right) \right\|_{BC^k(\mathbf{R}^n, B(\mathbf{C}^m))}$$

$$\leq K \sup_{x \in \mathbf{R}^n} \left[\sup_{1 \leq i,j \leq m} \sup_{1 \leq |\alpha| \leq k} \left(1 + \sum_{\ell} |D^\alpha p_{i,j,\ell}(x)| \right)^k \cdot \right.$$

$$\left. \cdot \left(\sup_{|\alpha| \leq k} |D^\alpha g(x)| \right) \| \prod_{\ell} e^{(p_{i,j,\ell}(x))} \| \right) \right],$$

for any finite sequence $\{(p_{i,j,\ell})_{i,j=1}^m\}_\ell, g \in C^k(\mathbf{R}^n)$.

Proof of Theorem 30.2: Without loss of generality, we may assume that $\mu = 0, \omega < 0$.

Fix finite sequences $\{t_\ell\}_\ell \subseteq [0,\infty)$, $\{(p_{i,j,\ell})_{i,j=1}^m\}_\ell \subseteq \mathcal{A}_M$. Let

$$\mathcal{G}(x) \equiv \prod_{\ell} e^{t_\ell(p_{i,j,\ell}(x))}(1 + |x|^2)^{-r},$$

for $x \in \mathbf{R}^n$.

We will obtain a uniform bound for the norm of $\mathcal{G}(B)$, independent of $\{t_\ell\}_\ell, \{(p_{i,j,\ell})_{i,j=1}^m\}_\ell$.

Letting $g(x) \equiv (1 + |x|^2)^{-\frac{k}{2}(N-1)}$, in Lemma 30.10, we have

$$\|\mathcal{G}(x)(1 + |x|^2)^{r - \frac{k}{2}(N-1)}\|_{BC^k(\mathbf{R}^n, B(\mathbf{C}^m))}$$

$$= \left\| \left((\prod_{\ell} e^{(p_{i,j,\ell}(x))_{i,j=1}^m}) g(x) \right) \right\|_{BC^k(\mathbf{R}^n, B(\mathbf{C}^m))}$$

$$\leq K \left(1 + \sum_{\ell} (t_\ell + t_\ell^k) M (1 + |x|^2)^{\frac{1}{2}(N-1)} \right)^k (1 + |x|^2)^{-\frac{k}{2}(N-1)} e^{\omega \sum_\ell t_\ell}$$

$$= \sum_{i=0}^{k} \binom{k}{i} (\sum_{\ell} (t_\ell + t_\ell^k)^i M^i (1 + |x|^2)^{\frac{1}{2}(N-1)(i-k)} e^{\omega \sum_\ell t_\ell}$$

$$\leq (1 + M^k) \sum_{i=0}^{k} \binom{k}{i} \left((\sum_{\ell} t_\ell) + (\sum_{\ell} t_\ell)^k \right)^i e^{\omega \sum_\ell t_\ell}$$

$$\leq K(\omega) \equiv (1 + M^k) \sup_{s \geq 0} \sum_{i=0}^{k} \binom{k}{i} (s + s^k)^i e^{\omega s}.$$

By Proposition 12.3,

$$\|\mathcal{G}(B)\| \le M(r - \frac{k}{2}(N-1))K(\omega).$$

Straightforward calculations, using dominated convergence and the fact that $f \mapsto f(B)$ is an algebra homomorphism, now imply that

$$\mathcal{G}(B) = \prod_\ell e^{t_\ell(p_{i,j,\ell}(B))}(1 + |B|^2)^{-r}.$$

Thus we have the desired existence family. ∎

Proof of Theorem 30.4: This is the same as the proof of Theorem 30.2, with $g(x) \equiv e^{-|x|}(1 + |x|^2)^k$.

Proof of Theorems 30.3 and 30.5: Since $\mathcal{A} = \bigcup_{0 \le M} \mathcal{A}_M$, this follows immediately from Theorems 30.2, 30.4 and 29.4.

Proof of Theorem 30.9: As with the proof of Theorem 30.4, for $\{t_\ell\}_\ell \subseteq [0, \infty)$, $\{(p_{i,j,\ell})_{i,j=1}^m\}_\ell$ arbitrary, let $\mathcal{G}(x) \equiv \left(\prod_\ell e^{t_\ell(p_{i,j,\ell}(x))}\right) e^{-|x|}$, for $x \in \mathbf{R}^n$. Then

$$\mathcal{G}(B) = \left(\prod_\ell e^{t_\ell(p_{i,j,\ell}(B))}\right) e^{-|B|},$$

giving us the desired existence family. ∎

XXXI. TIME DEPENDENT EVOLUTION EQUATIONS

In this chapter, we apply the simultaneous solution space and simultaneous existence families to treat time-dependent evolution equations,

$$\frac{d}{dt}u(t) = A(t)(u(t)),\, u(s) = x\, (0 \leq s \leq t \leq T), \qquad (31.1)$$

by letting $I \equiv [0,T], A_\alpha \equiv A(t)$. In dealing with (31.1), a common hypothesis is to have, for some real ω, $(A(t) - \omega)$ generate a strongly continuous semigroup of contractions, for each $0 \leq t \leq T$, in some equivalent norm (this is saying that $\{A(t)\}_{0 \leq t \leq T}$ is a *stable* family).

In this section, we show how (31.1), when not well-posed in its original formulation, that is, when there does not exist an evolution system for $\{A(t)\}_{0 \leq t \leq T}$, may be dealt with using the simultaneous solution space for $\{A(t)\}_{0 \leq t \leq T}$, $Z_\omega(\{A(t)\}_{0 \leq t \leq T})$. As an example, we apply this to very general time dependent, space independent partial differential initial value problems (31.5) (see Example 31.4).

In particular, we want to emphasize how the simultaneous solution space for $\{A(t)\}_{0 \leq t \leq T}$, or, equivalently, the existence of a simultaneous existence family, is the central concept in dealing with ill-posed versions of (31.1), and reduces them in a straightforward way to the well-posed case.

Theorem 31.2. *Suppose $\{A(t)\}_{0 \leq t \leq T}$ is a family of closed operators on X, C_1 and C_2 are commuting members of $B(X)$, such that*

(1) *the map $t \mapsto A(t)C_2$ is a continuous map from $[0,T]$ into $B(X)$;*

(2) *$C_i A(t) \subseteq A(t)C_i$, for $0 \leq t \leq T, i = 1, 2$;*

(3) *there exists a $O(e^{\omega|t|})$ simultaneous C_1-existence family for $\{A(t)\}_{0 \leq t \leq T}$; and*

(4) *$Im(C_i)$ is dense, for $i = 1, 2$.*

Then there exists a Banach space W and a strongly continuous evolution system $U(t,s)$ for $A(t)|_W$ such that

$$[Im(C_1)] \hookrightarrow W \hookrightarrow X,$$

and

$$\|U(t,s)\|_W \leq e^{\omega(t-s)}\, (0 \leq s \leq t \leq T).$$

Remarks 31.3. Note that (1) and (4) imply that $\bigcap_{0 \leq t \leq T} \mathcal{D}(A(t))$ is dense, thus (3) is equivalent, by Theorem 29.5, to

196

(3') $Im(C) \subseteq Im\left(\prod_{k=1}^{n}(\lambda_k - A(t_k))\right)$, for all finite sequences $\{\lambda_k\}_{k=1}^{n} \subseteq$ $(\omega, \infty), \{t_k\}_{k=1}^{n} \subseteq [0, T]$, and there exists a constant M such that

$$\left\| \left(\prod_{k=1}^{n} (\lambda_k - \omega)(\lambda_k - A(t_k))^{-1} \right) C_1 \right\| \leq M,$$

for all $\{\lambda_k\}_{k=1}^{n} \subseteq (\omega, \infty), \{t_k\}_{k=1}^{n} \subseteq [0, T], n - 1 \in \mathbf{N}$.

When $C = I$, this is saying that $\{A(t)\}$ is a *stable* family of operators.

In fact, $Z_\omega(\{A(t)\}_{0 \leq t \leq T})$ is the maximal continuously embedded Banach subspace on which $\{A(t)\}_{0 \leq t \leq T}$ is stable with stability constant ω.

Of course, $Im(C_2)$ could be replaced by an appropriate continuously embedded Banach space Y, such that $t \mapsto A(t)$ is a continuous map from $[0, T]$ into $B(Y, X)$. Since we are not going into time-dependent evolution equations in depth, we chose to present a simpler result to give a more pleasant illustration of our simultaneous existence families.

Of particular interest is the case when $\omega \leq 0$; then we have a dissipative evolution system on W.

Example 31.4. We may apply Theorem 31.2 to systems of time dependent, space independent partial differential initial value problems

$$\frac{d}{dt} \vec{u}(t, x) = (p_{i,j}(t, D))_{i,j=1}^{m} \vec{u}(t, x), \ \vec{u}(0, x) = \vec{g}(x) \ (x \in \mathbf{R}^n, 0 \leq t \leq T),$$

$$(31.5)$$

on $[L^p(\mathbf{R}^n)]^m (1 \leq p < \infty)$, or any space of functions on $\mathbf{R}^n$ where translation is strongly continuous and uniformly bounded, $D \equiv i(\frac{\partial}{\partial x_1}, \ldots, \frac{\partial}{\partial x_n})$ and for $0 \leq t \leq T, 1 \leq i, j \leq m, p_{i,j}(t, x)$, as a function of x, is a polynomial in n variables.

Consider the following hypotheses, for some N, ω.

(H1) The map $t \mapsto p_{i,j}(t, x)$, from $[0, T]$ into $\mathbf{C}$, is continuous, for all $x \in \mathbf{R}^n, 1 \leq i, j \leq m$.

(H2) $n.r.(p_{i,j}(t, x))_{i,j}^{m} \subseteq \{z \in \mathbf{C} \mid Re(z) \leq \omega\}$, for $0 \leq t \leq T, x \in \mathbf{R}^n$, where $n.r.$ is the *numerical range* of a complex-valued matrix (see Terminology 13.3).

(H3) There exists $M < \infty$ such that $|D^\alpha p_{i,j}(t, x)| \leq Me^{-|x|}$, for $0 \leq t \leq T, x \in \mathbf{R}^n, 1 \leq i, j \leq m, |\alpha| \leq [\frac{n}{2} + 1]$.

(H4) The degree of $p_{i,j}(t, x)$ is less than or equal to N, for $0 \leq t \leq T, 1 \leq i, j \leq m$.

Then the following theorem is a consequence of Theorems 31.2, 30.4 and 30.2, Proposition 12.3, and, for (H1), a short dominated convergence argument, with, for r, s as in Proposition 12.3 and Theorem 30.2, $C_1 \equiv (1 + \triangle)^{-r}$ for (b), $C_1 \equiv e^{-|D|}$ for (a), $C_2 \equiv e^{-|D|}(1 + \triangle)^{-s}$; for (b), note that (H4) and (H1) imply that there exists M such that $|D^\alpha p_{i,j}(t, x)| \leq M(1 + |x|^2)^{\frac{1}{2}(N - |\alpha|)}$, for $0 \leq t \leq T, x \in \mathbf{R}^n, 1 \leq i, j \leq m, |\alpha| \leq [\frac{n}{2} + 1]$.

Note that the degree of p_t could be unbounded, so that $(H4)$ does not hold; let $p(t, x) = \sum_{k=1}^\infty a_k(t) x^k$, where, for each k, a_k is continuous and supported on $[1 - \frac{1}{k}, 1], a_k(1) = 0$.

Theorem. Let $X = L^p(\mathbf{R}^n)(1 \leq p < \infty), C_0(\mathbf{R}^n)$, or any space of functions on $\mathbf{R}^n$ where translation, in any direction, is a strongly continuous uniformly bounded group of operators. Then

(a) under hypotheses (H1), (H2) and (H3), for any $\mu > \omega$, there exists a Banach space W_μ such that

$$[Im(e^{-|D|})]^m \hookrightarrow W_\mu \hookrightarrow X^m,$$

and (31.5) is well-posed on W_μ, that is, there exists an evolution system $U(t, s)$ for $\{(p_{i,j}(t, D))_{i,j=1}^m | w_\mu\}_{0 \leq t \leq T}$, with

$$\|U(t, s)\| \leq e^{\mu(t-s)} \ (0 \leq s \leq t \leq T);$$

(b) under hypotheses (H1), (H2) and (H4), for any $\mu > \omega, r > \frac{1}{2}[\frac{n}{2} + 1](N - 1) + \frac{n}{4}$, there exists a Banach space $W_{\mu,r}$ such that

$$[\mathcal{D}(\triangle^r)]^m \hookrightarrow W_{\mu,r} \hookrightarrow X^m,$$

and (31.5) is well-posed on $W_{\mu,r}$, that is, there exists an evolution system $U(t, s)$ for $(p_{i,j}(t, D))_{i,j=1}^m | w_\mu$, with

$$\|U(t, s)\| \leq e^{\mu(t-s)} \ (0 \leq s \leq t \leq T).$$

Proof of Theorem 31.2: By translating, we may assume, without loss of generality, that $\omega = 0$. Let $Z \equiv Z_0(\{A(t)\}_{0 \leq t \leq T})$. By (3) and Theorem 29.5,

$$[Im(C_1)] \hookrightarrow Z \hookrightarrow X. \tag{31.6}$$

Let $Y \equiv C_1 C_2(Z)$ and let W equal the closure, in Z, of Y.

To see that $[Im(C_1)] \hookrightarrow W$, it is sufficient, by (31.6), to show that $Im(C_1) \subseteq W$. So suppose $x \in X$. By (31.6), $C_2(Z)$ contains $Im(C_1 C_2)$, thus, by (4), there exists a sequence $< x_n > \subseteq Z$ such that $C_2 x_n \to x$, in X. Thus $C_1 C_2 x_n \to C_1 x$, in $[Im(C_1)]$, hence, by (31.6), in Z. Since $C_1 C_2 x_n \in Y$, for all n, $C_1 x \in W$.

We make Y a Banach space in the usual way:

$$\|y\|_Y \equiv \inf\{\|x\|_Z \mid C_1 C_2 x = y\}.$$

We will show the following.

(a) For $0 \leq t \leq T$, $A(t)|_W$ and $A(t)|_Y$ generate strongly continuous semigroups of contractions.

(b) The map $t \mapsto A(t)|_W$ is a continuous map from $[0, T]$ into $B(Y, W)$.

For (a), note that, by (2), $C_1 C_2$ takes Z to Z, and

$$e^{s A(t)|_Z} C_1 C_2 x = C_1 C_2 e^{s A(t)|_Z} x, \tag{31.7}$$

for all $x \in Z, s \geq 0, 0 \leq t \leq T$. Thus $e^{s A(t)|_Z}$ leaves Y invariant, hence leaves W invariant. For $0 \leq t \leq T$, this implies that $A(t)|_W$ generates a strongly continuous semigroup of contractions because $A(t)|_Z$ does. For $A(t)|_Y$, all that remains is to check, for $y \in Y$, $\|e^{s(A(t)|_Y)} y\|_Y$; by (31.7), this is less than or equal to $\|e^{s A(t)|_Z} x\|_Z \leq \|x\|_Z$, whenever $x \in Z$ and $C_1 C_2 x = y$, so that, by taking infima over all such x, we conclude that

$$\|e^{s(A(t)|_Y)} y\|_Y \leq \|y\|_Y,$$

for all $y \in Y, s \geq 0, 0 \leq t \leq T$, as desired.

For (b), note first that, for $x \in Z$, by (2), $A(t) C_1 C_2 x = C_1(A(t) C_2 x) \in Im(C_1)$, thus is contained in W, that is, $A(t) : Y \to W$.

By (1), for $0 \leq s, t \leq T$, there exists $M(s, t)$ such that

$$\|A(s) C_2 x - A(t) C_2 x\|_X \leq M(s, t)\|x\|_X,$$

for all $x \in X$, and $\lim_{s \to t} M(s, t) = 0$, for $0 \leq t \leq T$.

By (31.6), there exists $\delta > 0$ such that $\|x\|_Z \leq \delta \|x\|_{[Im(C_1)]}$, for all x in the image of C_1.

For $0 \leq s, t \leq T$, suppose $x \in Z$ and $y = C_1 C_2 x$. Then

$$
\begin{aligned}
\|A(s) y - A(t) y\|_W &= \|A(s) C_1 C_2 x - A(t) C_1 C_2 x\|_Z \\
&\leq \delta \|A(s) C_1 C_2 x - A(t) C_1 C_2 x\|_{[Im(C_1)]} \\
&\leq \delta \|A(s) C_2 x - A(t) C_2 x\|_X \\
&\leq \delta M(s, t)\|x\|_X \leq \delta M(s, t)\|x\|_Z.
\end{aligned}
$$

Taking infima on the right, over all $x \in Z$ such that $C_1 C_2 x = y$, gives us

$$\|A(s)y - A(t)y\|_W \le \delta M(s,t)\|y\|_Y,$$

for $0 \le s, t \le T, y \in Y$.

This proves (b). Clearly $Y \hookrightarrow W$ and is dense in W. Therefore, by a standard theorem for strongly continuous evolution systems (see, for example, [Paz, Thm. 3.1, page 135]), there exists a strongly continuous evolution system, U, for $\{A(t)|_W\}_{0 \le t \le T}$, satisfying the desired growth condition. ∎

NOTES

II.

This chapter is new.

Exponentially bounded existence families were introduced in [dL9] (see also [dL4] and [dL10]), defined via the Laplace transform

$$(r - A)^{-1}Cx = \int_0^\infty e^{-rt}W(t)x\,dt \quad (x \in X). \tag{*}$$

Definition (*), under appropriate hypotheses on $W(t)$, is shown in [dL9] to be equivalent to Definitions 2.3 and 2.4. This Laplace transform is clear when we think of the C-existence family $\{W(t)\}_{t \geq 0}$ as being $\{e^{tA}C\}_{t \geq 0}$. We prefer to use a more direct approach here, which is also more general in that it does not require exponential boundedness or the injectivity of $(r - A)$, for r large; a consequence of Theorem 2.6 and Lemma 2.10 is that our mild C-existence family, as defined here, when exponentially bounded, will satisfy the Laplace transform (*), for sufficiently large r, when $(r - A)$ is injective.

When C commutes with A, and $\{W(t)\}_{t \geq 0}$ is exponentially bounded, then $(r - A)$ is injective, for r sufficiently large (see Proposition 17.2).

We also introduced in [dL9], primarily as a tool for proving Theorem 17.7, an exponentially bounded *n-times integrated C-existence family* $\{S(t)\}_{t \geq 0}$, as a strongly continuous family of operators that satisfied the Laplace transform

$$(r - A)^{-1}Cx = r^n \int_0^\infty e^{-st}S(t)x\,dt \quad (x \in X),$$

for r large (see Chapter XVIII for integrated semigroups).

For a purely Laplace transform approach to (exponentially bounded solutions of) the abstract Cauchy problem, see [He-Neu] and [Neu5].

The idea of a semigroup of *unbounded* operators is introduced in [Hu]. The idea is to have a dense subset on which the semigroup property holds. To obtain anything analogous to well-posedness (see Theorem 2.6 and Chapters IV and V), one needs bounded operators. As indicated in Chapter I, having bounded operators to work with greatly enlarges the supply of mathematical armament. The definition of the generator, in [Hu], is somewhat indirect, involving the Laplace transform; the domain is essentially restricted to those x for which the map $t \mapsto T(t)x$ is exponentially

201

bounded. This generator may not be closed, and could have trivial domain (compare with Proposition 3.10).

One may think of a C-existence family as $T(t)C$, for a semigroup of unbounded operators $T(t)$, where C is chosen so that $T(t)C \in B(X)$, for all $t \geq 0$. It is an open question whether, given any semigroup of unbounded operators $\{T(t)\}_{t \geq 0}$, there exists $C \in B(X)$ such that $\{T(t)C\}_{t \geq 0}$ is a mild C-existence family. For a partial answer, see [dL13, section VI].

One may also consider the inhomogeneous abstract Cauchy problem, under the hypothesis that there exists a C-existence family for A (see [Ha-Kim-Kim] for a result of this sort, when C commutes with A). The simplest way to generate such results is to go down to a solution space for A (see Chapters IV and V), and use results for inhomogeneous abstract Cauchy problems, when A generates a strongly continuous semigroup.

It is clear what the analogous definition for a C-existence family for the second order abstract Cauchy problem (22.2) (which we think of as $\cos(itA^{\frac{1}{2}})C$) would be. More generally, one could consider existence families for the general integrodifferential equation

$$\frac{d}{dt}u(t) = A(u(t)) + (B * u)(t) \ (t \geq 0), \ u(0) = x, \qquad (**)$$

where $\{B(t)\}_{t \geq 0}$ is a family of closed operators and $*$ denotes convolution on $[0, \infty)$. Note that, by choosing $B(t) \equiv 1$, we obtain a general second order abstract Cauchy problem; more generally, by choosing B to be an operator-valued polynomial,

$$B(t) \equiv \sum_{k=0}^{n} t^k B_k,$$

(**) becomes an arbitrary higher order abstract Cauchy problem.

See [dL4] for an outline of existence families for (**), including Hille-Yosida type theorems. See also [A-Ke], [dL6], and [dL-K], for results related to (**).

III.

Regularized semigroups were introduced by Da Prato ([D1]). They were independently introduced by Davies and Pang ([Dav-P]), where they were called C-*semigroups*. There are many problems with the latter terminology; for example, C sounds like C_0, which is a shorthand for strongly

202

continuous; also, "C-semigroup" obscures the fact that a *specific* C is being presented, that is, it implies "C-regularized semigroup, for some C," instead of "C-regularized semigroup, for this specific C," which is more accurate and much more informative. Thus we have returned to the terminology of Da Prato.

The recent interest in the subject was started by the introduction of integrated semigroups (see Chapter XVIII) in [A1]. Thus the definition of generator given in [Dav-P] involves a Laplace transform (see (*) in the notes for Chapter II). Independently of [D1], we introduced the more direct definition of the generator in 3.1, involving differentiating at zero (this is the definition in [D1]), as is done to define the generator of a strongly continuous semigroup, in [dL13]. (We should mention here that [dL13] was submitted in 1987.) In [D1] and [Dav-P], it is assumed that the image of C is dense; in [Dav-P], it is assumed that the regularized semigroup is exponentially bounded. The generality of Chapter III first appeared in [dL13].

Some other contributers to regularized semigroups are Pang ([P]), Neubrander, Hieber and Holderrieth ([Hi-Ho-Neu]), Lumer [Lu], Miyadera and Tanaka ([T1]), Keyantuo ([Key]), Zheng and Lei ([Z-Lei1, 2, 3]), Li and Shaw ([Li], [Li-Sh1, 2]) and Ha, Kim and Kim ([Ha-Kim-Kim]).

Most of the work of Miyadera and Tanaka may be found in [T1], thus we are, for convenience, referring the reader to [T1] for these results. Their work is not in the same spirit as this book, and there is very little overlap between the material of this book and the work in [T1]. Much of their focus is on what they call the *complete infinitesimal generator*, G, which is defined as follows. If $\{W(t)\}_{t \geq 0}$ is a C-regularized semigroup, then denote by $\mathcal{D}_W$ the set of all $x \in Im(C)$ such that the map $t \mapsto W(t)C^{-1}x$ is differentiable at $t = 0$. It is clear that $\mathcal{D}_W$ is contained in the domain of the generator of $\{W(t)\}_{t \geq 0}$. Then G is defined to be the closure of $A|_{\mathcal{D}_W}$, where A is the generator of $\{W(t)\}_{t \geq 0}$. In many ways, this appears to be a difficult object to work with. It seems more like the generator of the semigroup of unbounded operators $\{C^{-1}W(t)\}_{t \geq 0}$; as we have commented before, the real power of the operator-theoretic approach to the abstract Cauchy problem relies on obtaining bounded operators to work with.

Theorems 3.4, 3.5 and 3.7 are from [dL13]. Analogues of Theorems 3.4, 3.5 and 3.7 may be found in [D1], [Dav-P] and [T1].

Proposition 3.9 appears in [dL7]. Proposition 3.10 is from [dL15]. Special cases of Proposition 3.11 and Corollary 3.12 first appear in [T1]. Theorems 3.13 and 3.14 are new.

In [Dav-P], the following appears.

Open Question. If $Im(C)$ is dense and A generates an exponentially

203

bounded C-regularized semigroup, is $C(\mathcal{D}(A))$ a core for A?

It is shown in [Dav-P] that the answer to this open question is "yes" when C is accretive, or when $(a, \infty) \subseteq \rho(A)$.

It is clear how to define a local regularized semigroup $\{W(t)\}_{0 \leq t \leq b}$, and get corresponding results for the local abstract Cauchy problem; see [T1].

It is also clear how to develop an analogous theory of C-regularized cosine functions, $C(t)C(s) = \frac{1}{2}C(C(t+s) + C(t-s)), C(0) = C$ (think of $C(t)$ as being $\cos(itA^{\frac{1}{2}})$), their generators, and their relationship to the second order abstract Cauchy problem (22.2); see [Li-Sh2]. See also [Hi1] for integrated cosine functions, analogous to integrated semigroups (see Chapter XVIII).

One may also define integrated regularized semigroups (see [T1], [Li] and [Li-Sh2]) and integrated regularized cosine functions (see [Li] and [Li-Sh2]). See Chapter XVIII for the definition of an integrated semigroup. An n-times integrated C-regularized semigroup generated by A corresponds to an $(r - A)^{-n}C$-regularized semigroup generated by A (see [dL9]).

IV.

Most of this chapter is from [dL17]. In [dL14], we constructed the solution space when the operator generates a regularized semigroup, as in 4.14 and 4.15.

A Frechet space that corresponds to the abstract Cauchy problem having a unique solution for all initial data in the domain of A^n, for some n, appears in [U].

In [Hen-Her], for an operator, A, on a Frechet space whose resolvent set contains $(0, \infty)$, a maximal subspace is constructed on which A generates an *equicontinuous* semigroup. This is essentially the same construction as that of Kantorovitz on Banach spaces ([K2]), and is really quite different from the results of this chapter (we remark that a locally equicontinuous strongly continuous semigroup on a Frechet space may have empty resolvent set). Our space is larger than appears in [Hen-Her], or in [K2] when X is a Banach space, because we are including all initial data for which the abstract Cauchy problem has a solution, whereas the spaces in [Hen-Her] and [K2] include only initial data for which the abstract Cauchy problem has a bounded solution. The spaces in [Hen-Her] and [K2] are maximal in an operator-theoretic sense, rather than the "pointwise" sense of this chapter.

What is surprising in our results in this chapter is that *all* initial data that yield solutions of the abstract Cauchy problem can be put together and made well-posed. The spaces constructed in [A-Neu-Sc], [Hen-Her], [K2] and [T1] include only initial data that yield exponentially bounded solutions of the abstract Cauchy problem. Once one leaves the setting of a generator of a strongly continuous semigroup on a Banach space, it is to be expected that there will be solutions of the abstract Cauchy problem that are not exponentially bounded. Example 4.10 has a simple example of an operator on a Banach space for which the abstract Cauchy problem has a unique mild solution, for all initial data in the domain of A, none of which are exponentially bounded.

We created the "bumpy translation space" in Example 4.17, in response to the following query from F. Neubrander: if an operator generates an exponentially bounded integrated semigroup (see Chapter XVIII), do all local solutions of the abstract Cauchy problem extend to a global solution? For any $\epsilon > 0$, by choosing, in Example 4.17, $f \in X \bigcap C^1([-\epsilon, \infty))$, that is not in $C^1([-\delta, \infty))$, for any $\delta > \epsilon$, we obtain a solution of the abstract Cauchy problem on $[0, \epsilon]$ that does not extend to any larger interval. The operator A, in Example 4.17, generates a bounded once-integrated semigroup. Thus the answer to the query is "no".

A nice treatment of Corollaries 4.11 and 4.12 (on a Banach space) may be found in [Neu1].

V.

The definitions, proofs and exposition of this chapter are new. Many of the statements of the theorems may be found in [dL-K].

The idea of a maximal Banach subspace on which an arbitrary operator on a Banach space generates a strongly continuous semigroup of contractions first appears in [K2], where it is called the *Hille-Yosida space*, and, independently, in [Kr-L-C]. In [dL-K], we show that the Hille-Yosida space contains all initial data for which the abstract Cauchy problem has a bounded uniformly continuous solution. In [K2], a different version of maximality, an operator-theoretic form, is shown. Constructions of similar Banach subspaces appear in [A-Neu-Sc], under the assumption that A generates an integrated semigroup, and in [dL14] and [T1], under the assumption that A generates an exponentially bounded regularized semigroup.

205

In [K2], Z_0 is constructed as follows (see also [Kr-L-C] and Theorem 28.6). For an operator, A, such that $(0,\infty) \subseteq \rho(A)$, define

$$\|x\|_Y \equiv \sup\{\| \left(\prod_k (\lambda_k - A)^{-1} \right) x\| \,|\, \lambda_k > 0\}.$$

The Banach subspace Y is then defined to be the set of all x such that $\|x\|_Y < \infty$, and Z_0 is defined to be the closure, in Y, of $\mathcal{D}(A|_Y)$.

It is not hard to show that this definition of Y is equivalent to the definition of this chapter, when $(0,\infty) \subseteq \rho(A)$. For the approach of this chapter and [dL-K], we do not need to assume that $\rho(A)$ is nonempty.

Many of the theorems of this chapter may be found in [dL-K], by a different method of proof, using the Laplace transform. Given an operator-valued function f, on $(0,\infty)$, in [dL-K] we construct a maximal Banach subspace which contains all x for which the map $s \mapsto f(s)x$ is a Laplace transform, in some sense. For the Hille-Yosida space, we would be considering $f(s) \equiv (s - A)^{-1}$. For the second order abstract Cauchy problem (22.2), we would choose $f(s) \equiv s(s^2 - A)^{-1}$ (see [Ci]); this is a maximal subspace on which A generates a cosine family of contractions. See also [Liz] for a generalization of [K2] and [Ci] to Volterra equations of convolution type. Although the construction in [dL-K] is more general than this chapter (it may be applied to scalar-type spectral operators and well-bounded operators), our method in this chapter is much simpler and more direct.

In [dL-K] we also construct maximal subspaces on which an operator generates a strongly continuous holomorphic semigroup of contractions, of a fixed angle between 0 and $\frac{\pi}{2}$.

The norm $\|x\|_{Z_\omega}$ is applied in [Hu, Definition 2.1], to the domain of a semigroup of unbounded operators $\{T_t\}_{t\geq 0}$. Note that, for x in this domain, $u(t,x) = T_t x$. For the construction in [Hu], the operators T_t need to be closed. If we defined a semigroup, with domain Z (as in Chapter IV), by $T_t x \equiv u(t,x)$, it is not clear if T_t would be closed or closable, even when A is closed.

Lemma 5.25, an "integrated form of Widder's theorem," is due to Arendt ([A2]).

Theorem 5.17 (b) $\leftrightarrow$ (d) is essentially in [T1]. Our result is somewhat sharper, in that the rate of growth of the regularized semigroup matches the rate of growth of the strongly continuous semigroup.

In [Dav-P], Section 6, it is stated that the construction in [Mi-Oh-Ok] may be used to construct an interpolating space Y satisfying (b) of Theorem 5.17, when $Im(C)$ is dense and A generates an exponentially bounded

206

C-regularized semigroup. However, the construction there required that $(\lambda - A)^{-1}$ leave $Im(C)$ invariant; the following example shows that this is not true, in general. Let

$$A \equiv \begin{bmatrix} 0 & C^{-1} \\ 0 & 0 \end{bmatrix}, \mathcal{D}(A) \equiv X \times Im(C).$$

Then A generates the exponentially bounded C-regularized semigroup

$$W(t) \equiv \begin{bmatrix} C & t \\ 0 & C \end{bmatrix},$$

and

$$(\lambda - A)^{-1} = \begin{bmatrix} \frac{1}{\lambda} & \frac{1}{\lambda^2}C^{-1} \\ 0 & \frac{1}{\lambda} \end{bmatrix},$$

with the same domain as A, for all $\lambda \neq 0$, which clearly does not leave $Im(C)$ invariant.

VI.

This chapter is from [dL14]. Theorem 6.1 is also in [Key], by a different proof.

Theorem 0.2, in [A-Neu-Sc], has a result similar to Theorem 6.1, for exponentially bounded n-times integrated semigroups of type $w > 0$. By choosing $C \equiv (\lambda - A)^{-n}$, Theorem 0.2, in [A-Neu-Sc], is a corollary of Theorem 6.1 . Also Theorem 0.1 in [A-Neu-Sc] follows from Proposition 6.4.

By choosing $C \equiv (\lambda - B)^{-k}$, and using Proposition 3.9 and the equivalence between generating a k-times integrated semigroup and generating a $(\lambda - B)^{-k}$-semigroup, Corollary 0.3, in [A-Neu-Sc], is a corollary of Theorem 6.6.

The equivalence of (a) and (b) is essentially in Theorem 6.6 in [T1].

VII.

Entire regularized groups were introduced in [dL7]. The approach of this chapter, with entire vectors, is new.

Entire vectors, or more generally, *analytic* vectors, have been used a great deal to generate simple sufficient conditions for an operator to generate a strongly continuous semigroup. For a group, one obtains necessary and sufficient conditions (see [Nel], [Ch1, 2], [Dav1], [Re-Sim]).

The Proposition of Remark 7.11 has appeared in many places, but so far as we know, our proof is new.

In [Au1, 2], entire vectors and entire regularized groups are used to show that all solutions of the backwards heat equation are entire. More generally, in [Au1, 2], it is shown that, whenever $-A$ generates a strongly continuous holomorphic semigroup, then all solutions of the abstract Cauchy problem are entire. In the languages of Chapters IV and VII, this is saying that $Z = \mathcal{E}(A)$, where Z is the solution space for A.

In [Au-E], $\mathcal{E}(A)$ is used to study entire solutions of time-dependent Cauchy problems. A special case of their results includes sufficient conditions for $\mathcal{E}(A)$ to be contained in $\mathcal{E}(A + B)$.

VIII.

Theorem 8.2 and Example 6.8 are in [dL7], by an entirely (pardon the pun) different proof. We explicitly construct the regularized group, as in Remark 8.5. Lemma 8.8 is in [dL16].

We have seen that the generator of a holomorphic strongly continuous semigroup generates a C-regularized group, with $Im(C)$ dense, while there exist generators of strongly continuous semigroups that do not generate a C-regularized group, for any C. This leads naturally to the following.

Open Question. Suppose B generates a differentiable semigroup (see [Paz, Section 2.4]). Does there exist C such that B generates a C-regularized group?

More generally, it would be of interest to know when the generator A, of a strongly continuous semigroup, generates a C-regularized group. This corresponds to the the abstract Cauchy problem being reversible, for x in $C(\mathcal{D}(A))$.

IX.

In [dL7], we consider (9.2) by formally writing down e^{tA}.

If B were a nonzero complex number, then a linear algebra calculation shows that

$$e^{zA} = \frac{1}{2} \begin{bmatrix} (e^{zB^{\frac{1}{2}}} + e^{-zB^{\frac{1}{2}}}) & B^{-\frac{1}{2}}(e^{zB^{\frac{1}{2}}} - e^{-zB^{\frac{1}{2}}}) \\ B^{\frac{1}{2}}(e^{zB^{\frac{1}{2}}} - e^{-zB^{\frac{1}{2}}}) & (e^{zB^{\frac{1}{2}}} + e^{-zB^{1/2}}) \end{bmatrix}. \qquad (*)$$

For A to generate a strongly continuous semigroup, one needs $\pm B^{\frac{1}{2}}$ to generate a strongly continuous semigroup. Hence, the standard hypothesis for A to generate a strongly continuous semigroup (on $\mathcal{D}(B^{\frac{1}{2}}) \times X$) is that B have a square root that generates a strongly continuous group (see [Go2, Section 2.7]).

Using $(*)$, we construct an entire regularized group $\{e^{zA}B^{\frac{1}{2}}e^{-(I+B)^{\frac{1}{2}}}\}_{z \in \mathbb{C}}$. More generally, in [dL7], using similar linear algebra calculations, we consider

$$B \equiv \begin{bmatrix} 0 & I \\ B_1 & B_2 \end{bmatrix}, \qquad (**)$$

so that $u' = Bu$ becomes a general second order abstract Cauchy problem

$$u'' = B_1 u + B_2 u',$$

and construct an entire C-regularized group $\{e^{zB}C\}_{z \in \mathbb{C}}$, for appropriate C.

When X is a Hilbert space, we may also consider (9.2) when B is self-adjoint. Using the spectral theorem, it is not hard to show that the operator A of (9.2) generates a strongly continuous semigroup if and only if B is bounded above. For general self-adjoint B, using either the spectral theorem and $(*)$, or entire vectors as in Chapter IX, it may be shown that A generates an entire $e^{-|B|}$-regularized group. See [dL-E], where we also consider B as in $(**)$, and apply our results to ill-posed problems in linear elasticity.

X.

This chapter is from [Bo-dL2].

XI.

This chapter is also from [Bo-dL2].

Theorem 11.1 essentially appears in many places, in many different mathematical languages: in [Sj], in the language of Fourier multipliers, in [Balab-E1], in the language of smooth distribution groups, and in [Hi2], in the language of integrated semigroups. We offer our proof as a much simpler and clearer approach, showing how the family of operators generated by the Schrödinger operator is a special case of a "boundary value," as in Chapter X.

Theorem 11.4 improves some of the results in [Balab-E2] and [P], with a shorter proof.

XII.

This chapter is new.

See [Ste] for Lemma 12.10. See [Arv] for (12.2) ; see [Dav1] for $n = 1$.

A more general class of operators than generators of bounded strongly continuous groups is generators of n-times integrated groups (see Chapter XVIII) $\{S(t)\}_{t \in \mathbf{R}}$ that are $O(t^n)$. For such operators, a $C^\infty(\mathbf{R})$ functional calculus is constructed, in a completely different way than this chapter, in [Balab-E-J] and [J1], where they also show that such operators correspond to a certain type of spectral distribution. See also [J2] for a surprising result relating such operators to generators of strongly continuous holomorphic semigroups. It is shown in [J2] that, if iA generates an n-times integrated group $\{S(t)\}_{t \in \mathbf{R}}$ that is $O(t^n)$, and A has positive spectrum, then $-A$ generates a strongly continuous holomorphic semigroup of angle $\frac{\pi}{2}$; this implies that $-A^r$ also generates such a semigroup, for any $r > 0$ (see Corollary 24.3).

XIII.

This chapter is from [dL16].

Interesting general results on matrices of operators may be found in [N3], [En1], [En2] and [En-N].

In [Hi1] and [Hi3], when $A_k \equiv \frac{i\partial}{\partial x_k}$, on $L^p(\mathbf{R}^n)(1 < p < \infty)$, integrated semigroups are used to consider (13.1). One of the difficulties in applying integrated semigroups is that the resolvent set of the generator must be nonempty; for a matrix of operators, this is often not the case (see [N3]).

In [Hi-Ho-Neu], regularized semigroups are applied, to obtain a result similar to Theorem 13.9, when $A_k \equiv \frac{i\partial}{\partial x_k}$, on $L^p(\mathbf{R}^n)$ and some other

210

spaces, using different methods than this section, mainly, an analogue of Theorem 17.7 where C commutes with A.

Theorem 13.9 has recently been sharpened, in [Z-Lei2].

XIV.

This chapter is also from [dL16].

XV.

Example 15.1 is from [dL13]. Examples 15.2 and 15.3 are from [dL5].

XVI.

This chapter is from [dL9].

XVII.

Essentially the same results as Theorems 17.3, 17.4 and 17.7 may be found in [dL-K]. A version of Theorem 17.7, when C commutes with A, may be found in [Hi-Ho-Neu]. Analogues of Theorem 17.4 appear in [D1], [Dav-P] and [T1]. An integrated semigroup version of Theorem 17.7 may be found in [A2].

In [dL9], Theorem 17.7 is proven by immediately constructing a $(k+1)$-times integrated C-existence family for A (see the notes to Chapter II), $\{S(t)\}_{t \geq 0}$,

$$S(t)x \equiv \int_\Gamma e^{zt}(z-A)^{-1}Cx\,\frac{dz}{2\pi i z^{k+1}}.$$

This produces solutions of a $(k+2)$-times integrated abstract Cauchy problem (see [Hi-Ho-Neu], when C commutes with A); to get back to the abstract Cauchy problem (0.1), as in the conclusion of Theorem 17.7, a Laplace transform argument relating integrated C-existence families to C-existence families is required.

XVIII.

Integrated semigroups were introduced by Arendt ([A1]); see also [A2], [A3], [A-Ke], [A-Neu-Sc], [Hi1-3], [Hi-Ke], [Neu2-4], [Th], [Balab-E-J], [J1, 2] and [dL 5, 6].

When the family of operators is exponentially bounded, then it may be shown that the resolvent set of the generator contains a left half-plane, and may be defined with the Laplace transform (see [A2], [Neu2]); specifically, $\frac{1}{s^n}(s - A)^{-1}$ must be the Laplace transform of the n-times integrated semigroup.

Example 18.2 is from [dL17]. For exponentially bounded regularized semigroups and integrated semigroups, the equivalence of (d) and (e) in Theorem 18.3 appears in [T1] and [dL5]. Corollary 18.4 appeared in [dL5].

Some of the ideas behind integrated semigroups may be found in the work on distribution semigroups (see [Lio], [So], [Balab-E1]). We feel, though, that [A1] was ground breaking; the key idea is to obtain a strongly continuous function from $[0, \infty)$ into $B(X)$. Thus one may take the formal exponential e^{tA} (we think of the solution $u(t, x)$, of the abstract Cauchy problem, as being $e^{tA}x$), and either integrate n times, or smooth it with an operator C, that commutes with A, $W(t) \equiv Ce^{tA} = e^{tA}C$. We feel that, for most purposes, the latter technique is simpler, more general and more informative.

One advantage of regularized semigroups over integrated semigroups is that one has a grip on the classical solution, $u(t, Cx) = W(t)x$, of the abstract Cauchy problem, immediately. To get this from the n-times integrated semigroup, one must differentiate n times, thus it is harder to determine the behaviour of the classical solution by looking at the integrated semigroup. Algebraically, regularized semigroups are much easier to deal with. For example, if A_j generates a C_j-regularized semigroup, $\{W_j(t)\}_{t \geq 0}$, for $j = 1, 2$, and they commute with each other, then it is easy to show that an extension of $A_1 + A_2$ generates the C_1C_2-regularized semigroup $\{W_1(t)W_2(t)\}_{t \geq 0}$. This is even easier to believe, since we think of $W_j(t)$ as being $e^{tA_j}C_j$. If B_j, $j = 1, 2$, generate commuting n-times integrated semigroups, then it is not at all clear if $B_1 + B_2$ generates a k-times integrated semigroup, for any k, and if it did, it would be unpleasant to try to express this integrated semigroup in terms of the integrated semigroups generated by B_1 and B_2. However, it *is* easy to show that an extension of $B_1 + B_2$ generates a $(B_1 - r)^{-n}(B_2 - r)^{-n}$-regularized semigroup, for $r \in \rho(B_1) \cap \rho(B_2)$ (see Chapter XV). Requiring nonempty resolvent sets, as is done with integrated semigroups, can also be a problem, as with matrices of operators (see Chapters XIII and XIV). And of course, regularized semigroups are much more general, since an n-times integrated semigroup corresponds merely to choosing $C \equiv (r - A)^{-n}$, whereas, in general, C could be any bounded, injective operator.

The following, from [Hi-Ke] (see [Hi1] for a generalization), is a nice result that unifies first and second order abstract Cauchy problems. An operator A generates a cosine function on X if and only if $\mathcal{A} \equiv \begin{bmatrix} 0 & 1 \\ A & 0 \end{bmatrix}$ generates a once-integrated semigroup on $X \times X$.

XIX.

Exercise 19.1 may be found in [dL9], and, for C commuting with A and B, in [T1]. Theorem 19.2 is from [dL14]. Theorem 19.6 and Examples 19.7, 19.8, 19.9 and 19.10 are from [dL8]. Theorem 19.4 and Examples 19.5 and 19.11 are new. Some references for other multiplicative perturbations are [C-G], [Dor], [Gu-Lu] and [Gu-R] and, more recently, [Ho] and [dL-Neu]. Remaining in the class of generators of strongly continuous semigroups, under multiplicative perturbations, is a delicate matter.

Some generalizations of Theorem 19.6 are in [dL-J1], where we present a class of operators that is closed under commuting products.

XX.

The example given in the fourth paragraph is from [Gu-R], Example 1.2.

XXI.

Details of this chapter may be found in [dL11], or, for C commuting with A, in [D3], [T1] and [Z-Lei1].

Definition 21.2, when the domain of its generator is dense, may be shown ([T1]) to be equivalent to "holomorphic semigroups of class (H_n)" ([Ok]); see also [Oh].

Definition 21.3 first appeared in [D3].

Theorem 21.18 (a) $\leftrightarrow$ (b) is essentially in [Ok], where semigroups of class (H_n) are considered (see [T1]).

XXII.

This chapter is from [dL15].

We consider the idea of a *regularized functional calculus* (Definition 22.11) to be the fundamental idea for the operator theoretic approach to the abstract Cauchy problem. For the abstract Cauchy problem to have a mild solution, for all initial data x in the image of C, one wishes to define $e^{tA}C$. To discuss the behavior of these solutions, we need to understand the behavior of the map $f \mapsto f(A)C$; this is a regularized functional calculus.

Other results on regularized functional calculi may be found in [dL-E-J], [dL-J2], and [dL19, 22].

Results similar to Corollary 22.12, for operators of (-1)-type $S_{\frac{\pi}{2}}$ (that is, generators of exponentially decaying strongly continuous holomorphic semigroups) may be found in [Bo-dL1].

Integrated semigroup analogues of 22.22 and 22.30(b) may be found in [A-Ke].

We are indebted to J. R. Dorroh for showing us (22.33).

XXIII.

This chapter is also from [dL15].

XXIV.

This chapter is from [dL3].

Corollary 24.3 was first shown in [dL1], by a different method. There, we explicitly constructed the semigroup generated by $-A^n$, with an unbounded analogue of the Cauchy integral formula,

$$e^{-zA^n}x \equiv \int e^{-zw^n} (w - A)^{-1} x \, \frac{dw}{2\pi i} \quad (x \in X),$$

as is done to construct a strongly continuous holomorphic semigroup (see also Chapter XXII).

It is well known (see [Go2]) that the square of a generator of a strongly continuous group generates a strongly continuous holomorphic semigroup

214

of angle $\frac{\pi}{2}$ (this is the special case of Theorem 11, when $p(t) = t^2$). The special case of Corollary 24.3, when $p(t) = t^2$, appears in [Go1]. In that paper, it is also shown that, if $-A$ generates a cosine function, then $-A^{2n}$ generates a C_0 holomorphic semigroup of angle $\frac{\pi}{2}$, for $n = 1, 2, \ldots$.

XXV.

This chapter is from [Go-dL-S].

The iterated form of the abstract Cauchy problem (25.1) was introduced in [S]. In [S], (25.1) was shown to be well-posed, in a certain sense, if and only if A_k generates a strongly continuous semigroup for $1 \leq k \leq n$ (see also [Go-S, 1-4]). The d'Alembert formula of Theorem 25.4 was introduced in [Go-S4]. This was extended to the setting of integrated semigroups in [Go-Hi] (see Example 4.4 of [Go-Hi]). The results of this chapter and Chapter XXVI require regularized semigroups; neither integrated semigroups nor strongly continuous semigroups will suffice.

One may also consider *incomplete* (this means that the number of initial data is less than the order of the differential equation) abstract Cauchy problems of arbitrary order, by factoring as in (25.1) to obtain an incomplete iterated abstract Cauchy problem; see [dL12].

XXVI.

This chapter, and the following notes for this chapter, are also from [Go-dL-S].

The abstract framework for equipartition of energy was introduced in [Go-S2]. Let X be a complex Hilbert space. Let $H = H^*$ be self-adjoint on X. A *finite projection system* on X is $\{P_1, \ldots, P_N\}$ where $N \geq 2, P_i$ is a nonzero orthogonal projection and $\sum_{i=1}^{N} P_i = I$. The Schrödinger equation

$$i\frac{du(t)}{dt} = Hu(t) \, (t \geq 0) \tag{1}$$

admits *equipartition of energy* if, for some finite projection system $\{P_1, \ldots, P_N\}$, the jth partial energy

$$E_j(t) = \|P_j \exp\{-itH\}f\|^2$$

satisfies

$$E_j(t) \to c_j\|f\|^2$$

as $t \to \infty$ where $c_j > 0$ and $\sum_{j=1}^{N} c_j = 1$; this must hold for all $f \in X$, and c_j is independent of f.

The classical example is the abstract wave equation

$$w''(t) + A^2 w(t) = 0$$

where A is self-adjoint and injective on Y. One takes $X = Y \oplus Y$ and rewrites this in the form (1) with

$$u = \begin{bmatrix} w' \\ Aw \end{bmatrix}, \quad -iH = \begin{bmatrix} 0 & -A \\ A & 0 \end{bmatrix}.$$

Then $E_1(t) = \|w'(t)\|^2$ and $E_2(t) = \|Aw(t)\|^2$ are, respectively, the kinetic and potential energies. When A is spectrally absolutely continuous, equipartition of energy holds with $c_1 = c_2 = \frac{1}{2}$. For a generalization to equations of order $N = 2^M$ see [Go-S3]. Other results are in [Go-S4].

To show that an operator (or an equation) does not admit equipartition of energy is in principle merely a computation. But to show that an equation does not admit equipartition of energy at all requires showing that a computation fails to work for *all* finite projection systems. Not surprisingly then, there is no existing literature on necessary conditions for equipartition of energy.

XXVII–XXX.

This is from [dL20, 21]. See the notes to Chapters IV and V for the history of solution spaces for a single operator.

This approach unifies many areas of operator theory and evolution equations. In Chapter XXXI, we have seen how, by considering $I \equiv [0, T]$, $A_\alpha \equiv A(t)$, we may treat time-dependent evolution equations (31.1). By choosing $\{A_\alpha\}_\alpha = \{\pm i B^n\}_{n \in \mathbf{N}}$, for some closed operator B, we obtain a maximal continuously embedded Banach subspace on which B is densely defined and has a $C_0(\mathbf{R})$ functional calculus (see [dL21]). When X is reflexive, this subspace is the *semi-simplicity manifold* of Kantorovitz ([K1, p. 157]) and is the maximal continuously embedded Banach subspace on which B is a spectral operator of scalar type (see [dL18]).

XXXI.

The obvious time-dependent analogue of regularized semigroups is introduced in [T2], called a *C-regularized evolution system* for $\{A(t)\}_{0 \leq t \leq T}$, defined as follows.

$$U(t,s)U(s,r) = U(t,r)C(T \geq t \geq s \geq r \geq 0), \ U(t,t) = C(0 \leq t \leq T).$$

This definition is clear if you think of $U(t,s)$ as $e^{\int_s^t A(r)\,dr} C$.

What is of more interest is to have a continuously embedded Banach subspace on which (31.1) is well-posed, that is, on which $A(t)$ has a strongly continuous evolution system. It is straightforward to show, as in the following proposition, that this is at least as strong as having a C-regularized evolution system.

Proposition. *Suppose there exists injective $C \in B(X)$ and a Banach space W such that*

$$[Im(C)] \hookrightarrow W \hookrightarrow X$$

and $\{A(t)|_W\}_{0 \leq t \leq T}$ has an evolution system, U_W, that commutes with $C|_W$. Then $\{A(t)\}_{0 \leq t \leq T}$ has a C-regularized evolution system, given by $U(t,s) \equiv U_W(t,s)C$.

Of course, in practice, there is nothing desirable about having C injective; this is merely a limitation of C-regularized evolution systems that is avoided by the technique of this chapter. It is not at all clear what, if any, converse of this Proposition (analogous to Theorem 6.6) might be true. By using this proposition and the simultaneous solution space $Z_\omega(\{A(t)\}_{0 \leq t \leq T})$, results about C-regularized evolution systems may be quickly and simply reduced to the corresponding results for evolution systems (see Theorem 31.2 and its proof).

For a C-regularized evolution system analogue of (b) of the Theorem in Example 31.4, when $m = 1$, see [T2] and [Z-Lei2]; see also the Proposition above.

As mentioned in the Notes for Chapter VII, entire solutions of (31.1) are considered in [Au-E].

BIBLIOGRAPHY

[A1] W. Arendt, *Resolvent positive operators*, Proc. London Math. Soc. 54 (1987), 321–349.

[A2] W. Arendt, *Vector-valued Laplace transforms and Cauchy problems*, Israel J. Math. 59 (1987), 327–352.

[A3] W. Arendt, *Sobolev imbeddings and integrated semigroups*, in: 2nd Internat. Conference on Trends in Semigroup Theory and Evol. Equations, Delft 1989. L. N. in Pure and Appl. Math. 135, Marcel Dekker (1991), 29–40.

[A-Ke] W. Arendt and H. Kellermann, *Integrated solutions of Volterra integro-differential equations and applications*, Integro-differential Equations, Proc. Conf. Trento (1987), G. Da Prato and M. Ianelli (eds.), Pitman, 1989.

[A-Neu-Sc] W. Arendt, F. Neubrander, and U. Schlotterbeck, *Interpolation of semigroups and integrated semigroups*, Semigroup Forum 45 (1992), 26–37.

[Arv] W. B. Arveson, *On groups of automorphisms of operator algebras*, J. Func. An. 15 (1974), 217–243.

[Au1] L. Autret, *Vecteurs entiers et réversibilité des problèmes paraboliques*, thesis, Université de Poitiers, 1992.

[Au2] L. Autret, *Entire vectors and time reversible Cauchy problems*, Semigroup Forum 46 (1993), 347–351.

[Au-E] L. Autret and H. Emamirad, *Entire propagator*, Proc. Amer. Math. Soc., to appear.

[B-C] J. B. Baillon and Ph. Clement, *Examples of unbounded imaginary powers of operators*, J. Func. An. 100 (1991), 419–434.

[Balab] M. Balabane, *Quelques propositions pour un calcul symbolique*, These d'Etat, Université de Paris 7, 1982.

[Balab-E1] M. Balabane and H. Emamirad, *Smooth distribution semigroup and Schrödinger equation in $L^p(\mathbf{R}^n)$*, J. Math. An. Appl. 70 (1979), 61–71.

[Balab-E2] M. Balabane and H. Emamirad, *L^p estimates for Schrödinger evolution equations*, Trans. Amer. Math. Soc. 292 (1985), 357–373.

[Balab-E-J] M. Balabane, H. Emamirad and M. Jazar, *Spectral distributions and generalization of Stone's theorem*, Act. Appl. Math., to appear.

[Balak] A.V. Balakrishnan, *Fractional powers of closed operators and the semigroups generated by them*, Pac. J. Math. 10 (1960), 419–437.

[Bar-Go] P. Baras and J. A. Goldstein, *The heat equation with a singular potential*, Trans. Amer. Math. Soc. 284 (1984), 121–139.

218

[Be1] R. Beals, *On the abstract Cauchy problem*, J. Func. An. 10 (1972), 281–299.

[Be2] R. Beals, *Semigroups and abstract Gevrey spaces*, J. Func. An. 10 (1972), 300–308.

[Bo-dL1] K. Boyadzhiev and R. deLaubenfels, H^∞-functional calculus for perturbations of generators of holomorphic semigroups, Houston J. Math. 17 (1991), 131–147.

[Bo-dL2] K. Boyadzhiev and R. deLaubenfels, *Boundary values of holomorphic semigroups*, Proc. Amer. Math. Soc. 118 (1993), 113–118.

[Bo-dL3] K. Boyadzhiev and R. deLaubenfels, *Semigroups and resolvents of bounded variation, imaginary powers and H^∞ functional calculus*, Semigroup Forum 45 (1992), 372–384.

[Bu1] R. Bürger, *Perturbations of positive semigroups and applications to population genetics*, Math. Z. 197 (1988), 259–272.

[Bu2] R. Bürger, *Unbounded one-parameter semigroups, Fréchet spaces and applications*, in: 2nd Internat. Conference on Trends in Semigroup Theory and Evol. Equations, Delft 1989. L. N. in Pure and Appl. Math. 135, Marcel Dekker (1991), 29–40.

[C-G] B. Calvert and K. Gustafson, *Multiplicative perturbations of nonlinear m-accretive operators*, J. Func. An. 10 (1972), 149–158.

[Ch1] P. R. Chernoff, *Some remarks on quasi-analytic vectors*, Trans. Amer. Math. Soc. 167 (1972), 105–113.

[Ch2] P. R. Chernoff, *Quasi-analytic vectors and quasi-analytic functions*, Bull. Amer. Math. Soc. 81 (1975), 637–646.

[Ci] I. Cioranescu, *On the second order Cauchy's problem associated with a linear operator*, J. Math. An. and Appl. 154 (1991), 238–243.

[Cr-Paz-T] M. G. Crandall, A. Pazy, and L. Tartar, *Remarks on generators of analytic semigroups*, Israel J. Math. 32 (1979), 363–374.

[D1] G. Da Prato, *Semigruppi regolarizzabili*, Ricerche Mat. 15 (1966), 223–248.

[D2] G. Da Prato, *Semigruppi di crescenza n.*, Ann. Scuola Norm. Sup. Pisa 20 (1966), 753–782.

[D3] G. Da Prato, *R-semigruppi analitici ed equazioni di evoluzione in L^p*, Ricerche Mat. 16 (1967), 233–249.

[Dav1] E. B. Davies, "One-Parameter Semigroups," Academic Press, London, 1980.

[Dav2] E. B. Davies, *Kernel estimates for functions of second order elliptic operators*, Quart. J. Math. Oxford (2) 39 (1988), 37–46.

[Dav-P] E. B. Davies and M. M. Pang, *The Cauchy problem and a generalization of the Hille-Yosida theorem*, Proc. London Math. Soc. 55 (1987), 181–208.

[dL1] R. deLaubenfels, *Powers of generators of holomorphic semigroups,* Proc. Amer. Math. Soc. 99 (1987), 105–108.

[dL2] R. deLaubenfels, *Totally accretive operators,* Proceedings of the American Mathematical Society 103 (1988), 551–556.

[dL3] R. deLaubenfels, *Polynomials of generators of integrated semigroups,* Proceedings of the American Mathematical Society 107 (1989), 197–204.

[dL4] R. deLaubenfels, *C-existence families and improperly posed problems,* Semesterbericht Funktionalanalysis, Tübingen Wintersemester 1989/1990.

[dL5] R. deLaubenfels, *Integrated semigroups, C-semigroups and the abstract Cauchy problem,* Semigroup Forum 41 (1990), 83–95.

[dL6] R. deLaubenfels, *Integrated semigroups and integrodifferential equations,* Mathematische Zeitschrift 204 (1990), 501–514.

[dL7] R. deLaubenfels, *Entire solutions of the abstract Cauchy problem,* Semigroup Forum 42 (1991), 83–105.

[dL8] R. deLaubenfels *Bounded commuting multiplicative perturbations of generators of strongly continuous groups,* Houston Math. J. 17 (1991), 299–310.

[dL9] R. deLaubenfels *Existence and uniqueness families for the abstract Cauchy problem,* J. London Math. Soc. 44 (1991), 310–338.

[dL10] R. deLaubenfels, *C-existence families,* in: 2nd Internat. Conference on Trends in Semigroup Theory and Evol. Equations, Delft 1989. L. N. in Pure and Appl. Math. 135, Marcel Dekker (1991), 295–309.

[dL11] R. deLaubenfels, *Holomorphic C-existence families,* Tokyo Math. J. 15 (1992), 17–38.

[dL12] R. deLaubenfels, *Incomplete iterated Cauchy problems,* J. Math. An. and Appl. 168 (1992), 552–579.

[dL13] R. deLaubenfels, *C-semigroups and the Cauchy problem,* J. Func. An. 111 (1993), 44–61.

[dL14] R. deLaubenfels, *C-semigroups and strongly continuous semigroups,* Israel J. Math. 81 (1993), 227–255.

[dL15] R. deLaubenfels, *Unbounded holomorphic functional calculus and abstract Cauchy problems for operators with polynomially bounded resolvent,* J. Func. An. 114 (1993), 348–394.

[dL16] R. deLaubenfels, *Matrices of operators and regularized semigroups,* Math. Z. 212 (1993), 619–629.

[dL17] R. deLaubenfels, *Automatic well-posedness with the abstract Cauchy problem,* J. London Math. Soc., to appear.

[dL18] R. deLaubenfels, *Scalar operators and the semi-simplicity manifold,* Proc. Amer. Math. Soc. (accepted for publication).

[dL19] R. deLaubenfels, *Boundary values of holomorphic semigroups, H^∞ functional calculi, and the inhomogeneous abstract Cauchy problem*, Differential Equations in Banach Spaces, Proceedings of Conference in Bologna 1991, Lecture Notes in Pure and Applied Mathematics, Vol 148, Marcel-Dekker (1993), 181–194.

[dL20] R. deLaubenfels, *Simultaneous well-posedness*, in: 3rd Internat. Conference on Trends in Semigroup Theory and Evol. Equations, Han-sur-Lesse 1991, Marcel Dekker (1993), 101–115.

[dL21] R. deLaubenfels, *Simultaneous solution spaces and existence families for families of operators*, submitted.

[dL22] R. deLaubenfels, *Regularized functional calculi and evolution equations*, submitted.

[dL-E] R. deLaubenfels and H. Emamirad, *Application de la théorie des semi-groupes C-régularisés en élasticité linéaire*, C. R. Acad. Sci. Paris 316, Sér. I (1993), 759–762.

[dL-E-J] R. deLaubenfels, H. Emamirad and M. Jazar, *Spectral distributions and regularized scalar operators*, submitted.

[dL-J1] R. deLaubenfels and M. Jazar, *Commuting multiplicative perturbations*, Houston J. Math., to appear.

[dL-J2] R. deLaubenfels and M. Jazar, *Regularized spectral distributions*, submitted.

[dL-K] R. deLaubenfels and S. Kantorovitz, *Laplace and Laplace-Stieltjes spaces*, J. Func. An. 116 (1993), 1–61.

[Do-V1] G. Dore and A. Venni, *On the closedness of the sum of two closed operators*, Math. Z. 196 (1987), 189–201.

[Do-V2] G. Dore and A. Venni, *Some results about complex powers of closed operators*, J. Math. An. and Appl. 149 (1990), 124–136.

[Dor] J. R. Dorroh, *Contraction semigroups in a function space*, Pac. J. Math. 19 (1966), 35–38.

[Du-S1] N. Dunford and J. T. Schwartz, "Linear Operators," Part I, Interscience, New York (1958).

[Du-S2] N. Dunford and J. T. Schwartz, "Linear Operators," Part III, Interscience, New York (1971).

[Duo1] X. T. Duong, *H^∞ functional calculus of elliptic operators with C^∞ coefficients on L^p spaces of smooth domains*, J. Austral. Math. Soc. (Ser. A) 48 (1990), 113–123.

[Duo2] X. T. Duong, *H_∞ functional calculus of second order elliptic PDE on L^p spaces*, Miniconference on Operators in Analysis, Proc. of the Center for Math. Analysis, ANU, Canberra, 24 (1989), 91–102.

[E-J] H. Emamirad and M. Jazar, *Applications of spectral distributions to some Cauchy problems in $L^p(\mathbf{R}^n)$*, in: 2nd Internat. Conference on

Trends in Semigroup Theory and Evol. Equations, Delft 1989. L. N. in Pure and Appl. Math. 135, Marcel Dekker (1991), 143–152.

[En1] K. J. Engel, *Polynomial operator matrices*, Dissertation Tübingen (1988).

[En2] K. J. Engel, *Polynomial operator matrices as semigroup generators: the general case*, Int. Eq. and Operator Theory 13 (1990), 175–192.

[En-N] K. J. Engel and R. Nagel, *Cauchy problems for polynomial operator matrices on abstract energy spaces*, Forum Math. 2 (1990), 89–102.

[F1] H. O. Fattorini, "The Abstract Cauchy Problem," Addison Wesley, Reading, Mass., 1983.

[F2] H. O. Fattorini, *Ordinary differential equations in linear topological spaces, I*, J. Diff. Eqns. 5 (1969), 72–105.

[Fr] A. Friedman, "Generalized Functions and Partial Differential Equations," Prentice Hall, 1963.

[G-S] I. M. Gelfand and G. E. Shilov, "Generalized Functions," Vol. 3, Academic Press, New York, 1968.

[Gi-S] D. Gilliam and J. R. Schullenberger, *A class of symmetric hyperbolic systems with special properties*, Comm. Part. Diff. Equations 4 (1979), 509–536.

[Go1] J. A. Goldstein, *Some remarks on infinitesimal generators of analytic semigroups*, Proc. Amer. Math. Soc. 22 (1969), 91–93.

[Go2] J. A. Goldstein, "Semigroups of Operators and Applications," Oxford, New York, 1985.

[Go-dL-S] J. A. Goldstein, R. deLaubenfels and J. T. Sandefur, *Regularized semigroups, iterated Cauchy problems and equipartition of energy*, Monat. Math. 115 (1993), 47–66.

[Go-Hi] J. A. Goldstein and M. Hieber, *Scattering theory and equipartition of energy*, Semesterbericht Funktionalanalysis Tübingen (1990), 95–110.

[Go-S1] J. A. Goldstein and J. T. Sandefur, *Equipartition of energy for symmetric hyperbolic systems*, in "Constructive Approaches to Mathematical Models," ed. by C. Coffman and G. Fix, Academic Press, New York (1979), 395–411.

[Go-S2] J. A. Goldstein and J. T. Sandefur, *Abstract equipartition of energy theorems*, J. Math. Anal. Appl. 67 (1979), 58–74.

[Go-S3] J. A. Goldstein and J. T. Sandefur, *Equipartition of energy for higher order hyperbolic equations*, Comm. P. D. E. 7 (1982), 1217–1251.

[Go-S4] J. A. Goldstein and J. T. Sandefur, *An abstract D'Alembert formula*, SIAM J. Math. An. 18 (1987), 842–856.

[Go-Sv] J. A. Goldstein and R. Svirsky, *On a domain characterization of Schrödinger operators with magnetic vector potentials and singular potentials*, Proc. Amer. Math. Soc. 105 (1989), 317–323.

222

[Gu-Lu] K. Gustafson and G. Lumer, *Multiplicative perturbation of semi-group generators*, Pac. J. Math. 41 (1972), 731–742.

[Gu-R] K. Gustafson and D. Rao, *Numerical range and accretivity of operator products*, J. Math. An. Appl. 60 (1977), 693–702.

·Kim-Kim] K. S. Ha, J. H. Kim and J. K. Kim, *Linear abstract Cauchy problem associated with an exponentially bounded C-semigroup in a Banach space*, Bull. Korean Math. Soc. 27 (1990), 157–164.

[H-V] R. Hempel and J. Voigt, *On the L_p-spectrum of Schrödinger operators*, J. Math. Anal. Appl. 121 (1987), 138–159.

[He-Neu] B. Hennig and F. Neubrander, *On representations, inversions and approximations of Laplace transforms in Banach spaces*, Semester-bericht Funktionalanalysis, Tübingen (1990).

[Hen-Her] H. Henriquez and E. Hernandez, *On the abstract Cauchy problem in Frechet spaces*, Proc. Amer. Math. Soc. 115 (1992), 353–360.

[Hi1] M. Hieber, *Integrated semigroups and differential operators on L^p*, Dissertation, Universität Tübingen, 1989.

[Hi2] M. Hieber, *Integrated semigroups and differential operators in L^p spaces*, Math. Ann. 291 (1991), 1–16.

[Hi3] M. Hieber, *Integrated semigroups and the Cauchy problem for systems in L^p spaces*, J. Math. An. Appl. 162 (1991), 300–308.

Ii-Ho-Neu] M. Hieber, A. Holderrieth and F. Neubrander, *Regularized semigroups and systems of linear partial differential equations*, Ann. Scuola Norm. di Pisa 19 (1992), 363–379.

[Hi-Ke] M. Hieber and H. Kellermann, *Integrated semigroups*, J. Func. An. 84 (1989), 160–180.

[Hil] E. Hille, *Une généralisation du problème de Cauchy*, Ann. Inst. Fourier 4 (1952), 31–48.

[Hil-P] E. Hille and R. S. Phillips, "Functional Analysis and Semigroups," Colloq. Publ. Amer. Math. Soc., 1957.

[Ho] A. Holderrieth, *Commuting multiplicative perturbations*, in: 3rd Internat. Conference on Trends in Semigroup Theory and Evol. Equations, Han-sur-Lesse 1992, Marcel Dekker, to appear.

[Hu] R. J. Hughes, *Semigroups of unbounded linear operators in Banach space*, Trans. Amer. Math. Soc. 230 (1977), 113–145.

[Hu-K] R. J. Hughes and S. Kantorovitz, *Boundary values of holomorphic semigroups of unbounded operators and similarity of certain perturbations*, J. Func. An. 29 (1978), 253–273.

[J1] M. Jazar, *Sur la théorie de la distribution spectrale et applications aux problèmes de Cauchy*, thesis, Université de Poitiers, 1991.

[J2] M. Jazar, *Fractional powers of momentum of a spectral distribution*, Proc. Amer. Math. Soc., to appear.

223

[K1] S. Kantorovitz "Spectral Theory of Banach Space Operators," Lecture Notes Math., Vol. 1012, Springer, Berlin-Heidelberg-New York (1983).

[K2] S. Kantorovitz *The Hille-Yosida space of an arbitrary operator*, J. Math. An. and Appl. 136 (1988), 107-111.

[Ka] T. Kato, "Perturbation Theory for Linear Operators," Springer, Berlin and New York, 1966; 2nd ed., 1976.

[Key] V. Keyantuo, *Interpolation et extrapolation des C-semi-groupes*, Publ. Math. de Besançon, September 1990.

[Ko] H. Komatsu, *Fractional powers of operators*, Pac. J. Math. 19 (1966), 285-346.

[Komu] T. Komura, *Semigroups of operators in locally convex spaces*, J. Func. An. 2 (1968), 258-296.

[Kr-L-C] S. G. Krein, G. I. Laptev and G. A. Cretkova, *On Hadamard correctness of the Cauchy problem for the equation of evolution*, Soviet Math. Dokl. 11 (1970), 763-766.

[Li] Y.-C. Li, *Integrated C-semigroups and C-cosine functions of operators on locally convex spaces*, Ph.D. dissertation, National Central University, 1991.

[Li-Sh1] Y.-C. Li and S.-Y. Shaw, *Representation formulas for C-semigroups*, Semigroup Forum, to appear.

[Li-Sh2] Y.-C. Li and S.-Y. Shaw, *On generators of integrated C-semigroups and C-cosine functions*, Semigroup Forum, to appear.

[Lio] J. L. Lions, *Semi-groupes de distributions*, Portugalae Math. 19 (1960), 141-164.

[Liz] C. Lizama, *On Volterra equations associated with a linear operator*, Proc. Amer. Math. Soc., to appear.

[Lu] G. Lumer, *Generalized evolution operators and (generalized) C-semigroups*, in: 2nd Internat. Conference on Trends in Semigroup Theory and Evol. Equations, Delft 1989. L. N. in Pure and Appl. Math. 135, Marcel Dekker (1991), 347-356.

[M] A. McIntosh, *Operators which have an H^∞ functional calculus*, Miniconference on Operator Theory and PDE, Proc. of the Center for Math. Analysis, ANU, Canberra, 14 (1986), 210-231.

[M-Y] A. McIntosh and A. Yagi, *Operators of type ω without a bounded H^∞- functional calculus*, Miniconference on Operators in Analysis 1989, Proc. of the Center for Math. Analysis, ANU, Canberra, 24 (1989).

[Mi-Oh-Ok] I. Miyadera, S. Oharu and N. Okazawa, *Generation theorems of semigroups of linear operators*, Publ. R.I.M.S. Kyoto Un. 8, No. 3 (1973), 509-555.

[N1] R. Nagel (ed.), "One-Parameter Semigroups of Positive Operators," Lect. Notes Math. 1184, 1986, Springer-Verlag.

[N2] R. Nagel, *Well-posedness and positivity for systems of linear evolution equations*, Conf. del Seminario di Matematica, Bari (1985).

[N3] R. Nagel, *Towards a "matrix theory" for unbounded operator matrices*, Math. Z. 201 (1989), 57–68.

[Nel] E. Nelson, *Analytic vectors*, Ann. of Math. 70 (1959), 572–615.

[Neu1] F. Neubrander, *Well-posedness of abstract Cauchy problems*, Semigroup Forum 29 (1984), 74–85.

[Neu2] F. Neubrander, *Integrated semigroups and their applications to the abstract Cauchy problem*, Pacific J. Math. 135 (1988), 111–157.

[Neu3] F. Neubrander, *Integrated semigroups and their application to complete second order problems*, Semigroup Forum 38 (1989), 233–251.

[Neu4] F. Neubrander, *Abstract elliptic operators, analytic interpolation semigroups, and Laplace transforms of analytic functions*, Semesterbericht Funktionalanalysis, Tübingen, Wintersemester 1988/89.

[Neu5] F. Neubrander, *The Laplace transform and semigroups*, in: 3rd Internat. Conference on Trends in Semigroup Theory and Evol. Equations, Han-sur-Lesse 1992, Marcel Dekker, to appear.

[Neu-Str] F. Neubrander and B. Straub *Fractional powers of operators with polynomially bounded resolvent*, Semesterbericht Funktionalanalysis, Tübingen, Wintersemester 1988/1989.

[Oh] S. Oharu, *Semigroups of linear operators in a Banach space*, Publ. R.I.M.S. Kyoto Univ. 7 (1971), 205–260.

[Ok] N. Okazawa, *Operator semigroups of class (D_n)*, Math. Japonicae 18 (1973), 33–51.

[P] M. M. Pang *Resolvent estimates for Schrödinger operators in $L^p(\mathbf{R}^n)$ and the theory of exponentially bounded C-semigroups*, Semigroup Forum 41 (1990), 97–114.

[Pay] L. E. Payne, "Improperly posed problems in partial differential equations," SIAM, Philadelphia, Pa., 1975.

[Paz] A. Pazy, "Semigroups of Linear Operators and Applications to Partial Differential Equations," Springer, New York, 1983.

[Pr-Soh] J. Prüss and H. Sohr, *On operators with bounded imaginary powers in Banach spaces*, Math. Z. 203 (1990), 429–452.

[Re-Sim] M. Reed and B. Simon, "Methods of Modern Mathematical Physics, II," Academic Press, New York (1975).

[Ri1] W. Ricker, *Spectral properties of the Laplace operator in $L^p(\mathbf{R})$*, Osaka J. Math., 25 (1988), 399–410.

[Ri2] W. Ricker, *An L^1-type functional calculus for the Laplace operator in $L^p(\mathbf{R})$*, J. Operator Theory, 21 (1989), 41–67.

[S]   J. T. Sandefur, *Higher order abstract Cauchy problems*, J. of Math. An. and Appl. 60 (1977), 728–742.

[Sau]   N. Sauer, *Linear evolution equations in two Banach spaces*, Proc. Roy. Soc. Edinburgh Sect. A 91 (1982), 287–303.

[Sau-Sin] N. Sauer and J. E. Singleton, *Evolution operators related to semigroups of class (A)*, Semigroup Forum 35 (1987), 317–335.

[Sim]   B. Simon, *Schrödinger semigroups*, Bull. Amer. Math. Soc. 7 (1982), 447–526.

[Sj]   S. Sjöstrand, *On the Riesz means of the solutions of the Schrödinger equation*, Annali Scuola Norm Sup. di Pisa 24 (1970), 331–348.

[So]   M. Sova, *Problèmes de Cauchy paraboliques abstraits de classes supérieurs et les semi-groupes de distributions*, Ricerche Mat. 18 (1969), 215–238.

[Ste]   E. M. Stein, "Singular Integrals and Differentiability Properties of Functions," Princeton University Press, New Jersey, 1970.

[Str]   B. Straub, *Über Generatoren von C-Halbgruppen und Cauchyprobleme zweiter Ordnung*, Diplomarbeit, Tübingen (1989).

[T1]   N. Tanaka, *C-semigroups of linear operators in Banach spaces—a generalization of the Hille-Yosida theorem*, thesis, Waseda University, 1992.

[T2]   N. Tanaka, *Linear evolution equations in Banach spaces*, Proc. London Math. Soc. 63 (1991), 657–672.

[Th]   H. R. Thieme, *Integrated semigroups and integrated solutions to abstract Cauchy problems*, J. Math. An. Appl. 152 (1990), 416–447.

[U]   T. Ushijima, *Some properties of regular distribution semigroups*, Proc. Japan Acad. 45 (1969) 224–227.

[vC]   J. A. van Casteren, "Generators of strongly continuous semigroups," Research Notes in Math., 115, Pitman, 1985.

[V]   A. Venni, *Some instances of the use of complex powers of linear operators*, Semesterbericht Funktionalanalysis, Tübingen, Sommersemester 1988, 235–246.

[Y1]   A. Yagi, *Coincidence entre des espaces d'interpolation et des domaines de puissances fractionnaires d'opérateurs*, C. R. Acad. Sci. Paris (Sér. I), 299 (1984), 173–176.

[Y2]   A. Yagi, *Applications of the purely imaginary powers of operators in Hilbert spaces*, J. Func. An. 73 (1987), 216–231.

[Yo]   K. Yosida, "Functional Analysis," Springer-Verlag, Berlin (1978).

[Z-Lei1] Q. Zheng and Y. Lei, *Exponentially bounded C-semigroups and integrated semigroups with nondensely defined generators I: Approximation; II: Perturbation; III: Analyticity*, Acta Math. Scientia, to appear.

226

[Z-Lei2] Q. Zheng and Y. Lei, *The application of C-semigroups to differential operators in $L^p(\mathbf{R}^n)$*, J. Math. An. and Appl., to appear.

[Z-Lei3] Q. Zheng and Y. Lei, *C-semigroups, integrated semigroups and integrated solutions to abstract Cauchy problem*, to appear.

SUBJECT INDEX

229

232

Springer-Verlag
and the Environment

We at Springer-Verlag firmly believe that an international science publisher has a special obligation to the environment, and our corporate policies consistently reflect this conviction.

We also expect our business partners – paper mills, printers, packaging manufacturers, etc. – to commit themselves to using environmentally friendly materials and production processes.

The paper in this book is made from low- or no-chlorine pulp and is acid free, in conformance with international standards for paper permanency.

Printing: Weihert-Druck GmbH, Darmstadt
Binding: Buchbinderei Schäffer, Grünstadt

Lecture Notes in Mathematics

For information about Vols. 1–1389
please contact your bookseller or Springer-Verlag

and Numerical Methods. Proceedings, 1988. VII, 238 pages. 1990.

Vol. 1432: K. Ambos-Spies, G.H. Müller, G.E. Sacks (Eds.), Recursion Theory Week. Proceedings, 1989. VI, 393 pages. 1990.

Vol. 1433: S. Lang, W. Cherry, Topics in Nevanlinna Theory. II, 174 pages.1990.

Vol. 1434: K. Nagasaka, E. Fouvry (Eds.), Analytic Number Theory. Proceedings, 1988. VI, 218 pages. 1990.

Vol. 1435: St. Ruscheweyh, E.B. Saff, L.C. Salinas, R.S. Varga (Eds.), Computational Methods and Function Theory. Proceedings, 1989. VI, 211 pages. 1990.

Vol. 1436: S. Xambó-Descamps (Ed.), Enumerative Geometry. Proceedings, 1987. V, 303 pages. 1990.

Vol. 1437: H. Inassaridze (Ed.), K-theory and Homological Algebra. Seminar, 1987–88. V, 313 pages. 1990.

Vol. 1438: P.G. Lemarié (Ed.) Les Ondelettes en 1989. Seminar. IV, 212 pages. 1990.

Vol. 1439: E. Bujalance, J.J. Etayo, J.M. Gamboa, G. Gromadzki. Automorphism Groups of Compact Bordered Klein Surfaces: A Combinatorial Approach. XIII, 201 pages. 1990.

Vol. 1440: P. Latiolais (Ed.), Topology and Combinatorial Groups Theory. Seminar, 1985–1988. VI, 207 pages. 1990.

Vol. 1441: M. Coornaert, T. Delzant, A. Papadopoulos. Géométrie et théorie des groupes. X, 165 pages. 1990.

Vol. 1442: L. Accardi, M. von Waldenfels (Eds.), Quantum Probability and Applications V. Proceedings, 1988. VI, 413 pages. 1990.

Vol. 1443: K.H. Dovermann, R. Schultz, Equivariant Surgery Theories and Their Periodicity Properties. VI, 227 pages. 1990.

Vol. 1444: H. Korezlioglu, A.S. Ustunel (Eds.), Stochastic Analysis and Related Topics VI. Proceedings, 1988. V, 268 pages. 1990.

Vol. 1445: F. Schulz, Regularity Theory for Quasilinear Elliptic Systems and – Monge Ampère Equations in Two Dimensions. XV, 123 pages. 1990.

Vol. 1446: Methods of Nonconvex Analysis. Seminar, 1989. Editor: A. Cellina. V, 206 pages. 1990.

Vol. 1447: J.-G. Labesse, J. Schwermer (Eds), Cohomology of Arithmetic Groups and Automorphic Forms. Proceedings, 1989. V, 358 pages. 1990.

Vol. 1448: S.K. Jain, S.R. López-Permouth (Eds.), Non-Commutative Ring Theory. Proceedings, 1989. V, 166 pages. 1990.

Vol. 1449: W. Odyniec, G. Lewicki, Minimal Projections in Banach Spaces. VIII, 168 pages. 1990.

Vol. 1450: H. Fujita, T. Ikebe, S.T. Kuroda (Eds.), Functional-Analytic Methods for Partial Differential Equations. Proceedings, 1989. VII, 252 pages. 1990.

Vol. 1451: L. Alvarez-Gaumé, E. Arbarello, C. De Concini, N.J. Hitchin, Global Geometry and Mathematical Physics. Montecatini Terme 1988. Seminar. Editors: M. Francaviglia, F. Gherardelli. IX, 197 pages. 1990.

Vol. 1452: E. Hlawka, R.F. Tichy (Eds.), Number-Theoretic Analysis. Seminar, 1988–89. V, 220 pages. 1990.

Vol. 1453: Yu.G. Borisovich, Yu.E. Gliklikh (Eds.), Global Analysis – Studies and Applications IV. V, 320 pages. 1990.

Vol. 1454: F. Baldassari, S. Bosch, B. Dwork (Eds.), p-adic Analysis. Proceedings, 1989. V, 382 pages. 1990.

Vol. 1455: J.-P. Françoise, R. Roussarie (Eds.), Bifurcations of Planar Vector Fields. Proceedings, 1989. VI, 396 pages. 1990.

Vol. 1456: L.G. Kovács (Ed.), Groups – Canberra 1989. Proceedings. XII, 198 pages. 1990.

Vol. 1457: O. Axelsson, L.Yu. Kolotilina (Eds.), Preconditioned Conjugate Gradient Methods. Proceedings, 1989. V, 196 pages. 1990.

Vol. 1458: R. Schaaf, Global Solution Branches of Two Point Boundary Value Problems. XIX, 141 pages. 1990.

Vol. 1459: D. Tiba, Optimal Control of Nonsmooth Distributed Parameter Systems. VII, 159 pages. 1990.

Vol. 1460: G. Toscani, V. Boffi, S. Rionero (Eds.), Mathematical Aspects of Fluid Plasma Dynamics. Proceedings, 1988. V, 221 pages. 1991.

Vol. 1461: R. Gorenflo, S. Vessella, Abel Integral Equations. VII, 215 pages. 1991.

Vol. 1462: D. Mond, J. Montaldi (Eds.), Singularity Theory and its Applications. Warwick 1989, Part I. VIII, 405 pages. 1991.

Vol. 1463: R. Roberts, I. Stewart (Eds.), Singularity Theory and its Applications. Warwick 1989, Part II. VIII, 322 pages. 1991.

Vol. 1464: D. L. Burkholder, E. Pardoux, A. Sznitman, Ecole d'Eté de Probabilités de Saint- Flour XIX-1989. Editor: P. L. Hennequin. VI, 256 pages. 1991.

Vol. 1465: G. David, Wavelets and Singular Integrals on Curves and Surfaces. X, 107 pages. 1991.

Vol. 1466: W. Banaszczyk, Additive Subgroups of Topological Vector Spaces. VII, 178 pages. 1991.

Vol. 1467: W. M. Schmidt, Diophantine Approximations and Diophantine Equations. VIII, 217 pages. 1991.

Vol. 1468: J. Noguchi, T. Ohsawa (Eds.), Prospects in Complex Geometry. Proceedings, 1989. VII, 421 pages. 1991.

Vol. 1469: J. Lindenstrauss, V. D. Milman (Eds.), Geometric Aspects of Functional Analysis. Seminar 1989-90. XI, 191 pages. 1991.

Vol. 1470: E. Odell, H. Rosenthal (Eds.), Functional Analysis. Proceedings, 1987-89. VII, 199 pages. 1991.

Vol. 1471: A. A. Panchishkin, Non-Archimedean L-Functions of Siegel and Hilbert Modular Forms. VII, 157 pages. 1991.

Vol. 1472: T. T. Nielsen, Bose Algebras: The Complex and Real Wave Representations. V, 132 pages. 1991.

Vol. 1473: Y. Hino, S. Murakami, T. Naito, Functional Differential Equations with Infinite Delay. X, 317 pages. 1991.

Vol. 1474: S. Jackowski, B. Oliver, K. Pawałowski (Eds.), Algebraic Topology, Poznań 1989. Proceedings. VIII, 397 pages. 1991.

Vol. 1475: S. Busenberg, M. Martelli (Eds.), Delay Differential Equations and Dynamical Systems. Proceedings, 1990. VIII, 249 pages. 1991.

Vol. 1476: M. Bekkali, Topics in Set Theory. VII, 120 pages. 1991.

Vol. 1477: R. Jajte, Strong Limit Theorems in Noncommutative L_2-Spaces. X, 113 pages. 1991.

Vol. 1478: M.-P. Malliavin (Ed.), Topics in Invariant Theory. Seminar 1989-1990. VI, 272 pages. 1991.

Vol. 1479: S. Bloch, I. Dolgachev, W. Fulton (Eds.), Algebraic Geometry. Proceedings, 1989. VII, 300 pages. 1991.

Vol. 1480: F. Dumortier, R. Roussarie, J. Sotomayor, H. Żołądek, Bifurcations of Planar Vector Fields: Nilpotent Singularities and Abelian Integrals. VIII, 226 pages. 1991.

Vol. 1481: D. Ferus, U. Pinkall, U. Simon, B. Wegner (Eds.), Global Differential Geometry and Global Analysis. Proceedings, 1991. VIII, 283 pages. 1991.

Vol. 1482: J. Chabrowski, The Dirichlet Problem with L^2-Boundary Data for Elliptic Linear Equations. VI, 173 pages. 1991.

Vol. 1483: E. Reithmeier, Periodic Solutions of Nonlinear Dynamical Systems. VI, 171 pages. 1991.

Vol. 1484: H. Delfs, Homology of Locally Semialgebraic Spaces. IX, 136 pages. 1991.

Vol. 1485: J. Azéma, P. A. Meyer, M. Yor (Eds.), Séminaire de Probabilités XXV. VIII, 440 pages. 1991.

Vol. 1486: L. Arnold, H. Crauel, J.-P. Eckmann (Eds.), Lyapunov Exponents. Proceedings, 1990. VIII, 365 pages. 1991.

Vol. 1487: E. Freitag, Singular Modular Forms and Theta Relations. VI, 172 pages. 1991.

Vol. 1488: A. Carboni, M. C. Pedicchio, G. Rosolini (Eds.), Category Theory. Proceedings, 1990. VII, 494 pages. 1991.

Vol. 1489: A. Mielke, Hamiltonian and Lagrangian Flows on Center Manifolds. X, 140 pages. 1991.

Vol. 1490: K. Metsch, Linear Spaces with Few Lines. XIII, 196 pages. 1991.

Vol. 1491: E. Lluis-Puebla, J.-L. Loday, H. Gillet, C. Soulé, V. Snaith, Higher Algebraic K-Theory: an overview. IX, 164 pages. 1992.

Vol. 1492: K. R. Wicks, Fractals and Hyperspaces. VIII, 168 pages. 1991.

Vol. 1493: E. Benoît (Ed.), Dynamic Bifurcations. Proceedings, Luminy 1990. VII, 219 pages. 1991.

Vol. 1494: M.-T. Cheng, X.-W. Zhou, D.-G. Deng (Eds.), Harmonic Analysis. Proceedings, 1988. IX, 226 pages. 1991.

Vol. 1495: J. M. Bony, G. Grubb, L. Hörmander, H. Komatsu, J. Sjöstrand, Microlocal Analysis and Applications. Montecatini Terme, 1989. Editors: L. Cattabriga, L. Rodino. VII, 349 pages. 1991.

Vol. 1496: C. Foias, B. Francis, J. W. Helton, H. Kwakernaak, J. B. Pearson, H_∞-Control Theory. Como, 1990. Editors: E. Mosca, L. Pandolfi. VII, 336 pages. 1991.

Vol. 1497: G. T. Herman, A. K. Louis, F. Natterer (Eds.), Mathematical Methods in Tomography. Proceedings 1990. X, 268 pages. 1991.

Vol. 1498: R. Lang, Spectral Theory of Random Schrödinger Operators. X, 125 pages. 1991.

Vol. 1499: K. Taira, Boundary Value Problems and Markov Processes. IX, 132 pages. 1991.

Vol. 1500: J.-P. Serre, Lie Algebras and Lie Groups. VII, 168 pages. 1992.

Vol. 1501: A. De Masi, E. Presutti, Mathematical Methods for Hydrodynamic Limits. IX, 196 pages. 1991.

Vol. 1502: C. Simpson, Asymptotic Behavior of Monodromy. V, 139 pages. 1991.

Vol. 1503: S. Shokranian, The Selberg-Arthur Trace Formula (Lectures by J. Arthur). VII, 97 pages. 1991.

Vol. 1504: J. Cheeger, M. Gromov, C. Okonek, P. Pansu, Geometric Topology: Recent Developments. Editors: P. de Bartolomeis, F. Tricerri. VII, 197 pages. 1991.

Vol. 1505: K. Kajitani, T. Nishitani, The Hyperbolic Cauchy Problem. VII, 168 pages. 1991.

Vol. 1506: A. Buium, Differential Algebraic Groups of Finite Dimension. XV, 145 pages. 1992.

Vol. 1507: K. Hulek, T. Peternell, M. Schneider, F.-O. Schreyer (Eds.), Complex Algebraic Varieties. Proceedings, 1990. VII, 179 pages. 1992.

Vol. 1508: M. Vuorinen (Ed.), Quasiconformal Space Mappings. A Collection of Surveys 1960-1990. IX, 148 pages. 1992.

Vol. 1509: J. Aguadé, M. Castellet, F. R. Cohen (Eds.), Algebraic Topology - Homotopy and Group Cohomology. Proceedings, 1990. X, 330 pages. 1992.

Vol. 1510: P. P. Kulish (Ed.), Quantum Groups. Proceedings, 1990. XII, 398 pages. 1992.

Vol. 1511: B. S. Yadav, D. Singh (Eds.), Functional Analysis and Operator Theory. Proceedings, 1990. VIII, 223 pages. 1992.

Vol. 1512: L. M. Adleman, M.-D. A. Huang, Primality Testing and Abelian Varieties Over Finite Fields. VII, 142 pages. 1992.

Vol. 1513: L. S. Block, W. A. Coppel, Dynamics in One Dimension. VIII, 249 pages. 1992.

Vol. 1514: U. Krengel, K. Richter, V. Warstat (Eds.), Ergodic Theory and Related Topics III, Proceedings, 1990. VIII, 236 pages. 1992.

Vol. 1515: E. Ballico, F. Catanese, C. Ciliberto (Eds.), Classification of Irregular Varieties. Proceedings, 1990. VII, 149 pages. 1992.

Vol. 1516: R. A. Lorentz, Multivariate Birkhoff Interpolation. IX, 192 pages. 1992.

Vol. 1517: K. Keimel, W. Roth, Ordered Cones and Approximation. VI, 134 pages. 1992.

Vol. 1518: H. Stichtenoth, M. A. Tsfasman (Eds.), Coding Theory and Algebraic Geometry. Proceedings, 1991. VIII, 223 pages. 1992.

Vol. 1519: M. W. Short, The Primitive Soluble Permutation Groups of Degree less than 256. IX, 145 pages. 1992.

Vol. 1520: Yu. G. Borisovich, Yu. E. Gliklikh (Eds.), Global Analysis – Studies and Applications V. VII, 284 pages. 1992.

Vol. 1521: S. Busenberg, B. Forte, H. K. Kuiken, Mathematical Modelling of Industrial Process. Bari, 1990. Editors: V. Capasso, A. Fasano. VII, 162 pages. 1992.

Vol. 1522: J.-M. Delort, F. B. I. Transformation. VII, 101 pages. 1992.

Vol. 1523: W. Xue, Rings with Morita Duality. X, 168 pages. 1992.

Vol. 1524: M. Coste, L. Mahé, M.-F. Roy (Eds.), Real Algebraic Geometry. Proceedings, 1991. VIII, 418 pages. 1992.